THE ILLUSTRATED ENCYCLOPEDIA OF

WILDLIFE

VOLUME 3

The Mammals

Part III

Wildlife Consultant

MARY CORLISS PEARL, Ph. D.

Distributed by Encyclopaedia Britannica
Educational Corporation

Grey Castle Press

Published by Grey Castle Press, 1991

Distributed by Encyclopaedia Britannica Educational Corporation, 1991

THE ILLUSTRATED ENCYCLOPEDIA OF WILDLIFE
Volume 3: THE MAMMALS—Part III

Copyright © EPIDEM-Istituto Geografico De Agostini S.p.A., Novara, Italy

Copyright © Orbis Publishing Ltd., London, England 1988/89

Organization and Americanization © Grey Castle Press, Inc., Lakeville, CT 06039

Library of Congress Cataloging-in-Publication Data
The Illustrated encyclopedia of wildlife.
     p.   cm.
     Contents: v. 1–5. The mammals — v. 6–8. The birds —
v. 9. Reptiles and amphibians — v. 10. The fishes —
v. 11–14. The invertebrates — v. 15. The invertebrates
and index.
     ISBN 1–55905–052–7
     1. Zoology.
QL45.2.I44 1991   90–3750
591—dc20   CIP

ISBN 1–55905–052–7 (complete set)
     1–55905–039–X (Volume 3)

Printed in Spain

Photo Credits
Photographs were supplied by *Archivio IGDA-P2*: 474; *Bruce Coleman*: 426, 452t, 478, 493, 495, 502, 525; (J. & D. Bartlett) 498b, 501, 521; (D. Bartlett) 431; (M. Boulton) 475, 492, 505; (J. Burton) 405b, 473, 479t, 479b, 487, 510, 522b, 523; (R.I.M. Campbell) 373; (R. Cartmell) 530; (A. Compost) 472; (G. Cubitt) 452b, 498t, 519b; (L.R. Dawson) 397t, 420; (A.J. Deane) 444; (J. Dermid) 409t; (F. Erize) 386b, 387, 405t, 411, 412, 439t, 451, 516b; (M.P.L. Fogden) 461; (F. Foott) 477; (F. Futil) 371; (M. Grant) 524t; (U. Hirsch) 385t; (P. Jackson) 456, 457, 470; (M.P. Kahl) 445, 493t; (LTD) 459b; (L. Lyon) 375, 379; (J. MacKinnon) 367; (N. Myers) 371b, 435, 453; (J. Pearson) 483; (J. Tory Peterson) 439b; (D. & M. Plage) 476; (G.D. Plage) 447, 458; (M. Qureshi) 459t, 485t, 529; (H. Reinhard) 363, 399b, 506, 507, 526, 527; (L. Lee Rue) 369t, 409c, 482; (L. Lee Rue III) 480; (G.B. Schaller) 471; (J. Simon) 419t; (Sullivan) 441; (Diana R. Sullivan) 449; (N. Tomalin) 399t, 519; (S. Trevor) 376, 469, 522t; (T. Van Wormer) 407, 433; (G. Ziesler) 372b; *E. Hosking*: 489; *Jacana*: (A. Antony) 511, 514; (A. Bertrand) 400, 509, 531; (J.P. Champraux) 513; (L. Chana) 377, 503; (M. Claye) 512; (A.R. Dezev) 391b; (Ernie) 402, 403; (Frederic) 372t, 393, 409b; (Gens) 494; (A. Kerneiss) 464, 465; (C. Klemm) 391t, 440; (Massart) 422, 433; (J. Prevost) 429; (A. Rainon) 463 (B. Rebouleau) 394; (J. Robert) 368, 401, 409, 430, 432, 442, 443, 446, 466, 485b, 488, 424b; (P. Varon) 528; (Varin-Visage) 499, 500, 508, 515, 516t, 517; *Okapia-Frankfurt*: 491; *Planet Earth Pictures*: (S. Avery) 496, 497; *Survival Anglia*: (L. & T. Bomford) 454, 455; (P. Summ) 381; (X. Sundance) 392; (Varin-Visage) 369b, 386t, 389, 395, 397b, 414, 415, 419b; (Zani-Vendal) 406; (Ziesler) 382, 383, 384, 385b, 413.

FRONT COVER: Hippopotamuses wallowing in Lake Manyara National Park, Tanzania (Bruce Coleman/Michael Freeman).

CONTENTS

SNOUTS, WARTS AND BRISTLES

Though not the most beautiful of creatures,
pigs and peccaries are sturdy and adaptable,
with a broad diet that includes plants
and small animals

White-lipped peccary

Babirusa

Bush pig

Wart hog

Collared peccary

Giant forest hog

The wild pigs and peccaries form two closely related families within the large group of animals that make up the even-toed ungulates. Though pigs and peccaries share many characteristics—they are all medium-sized mammals, with large heads, short necks, pronounced snouts and bristly bodies—the two families have entirely separate natural distributions. Peccaries are animals of the New World, while pigs belong to the Old World.

The pig family

The nine species of wild pig are found in Europe, Africa and Asia, in a range of habitats from tropical swamps to open grasslands. They are omnivorous animals, eating both plant and animal food, and their teeth are suited to their generalized diets. To cope with a variety of foods, their teeth are all-purpose tools able to grind, cut and shred. In many species, the canines are developed into long, upwardly curving tusks.

The faces of pigs are particularly distinctive. Their snouts are long enough to need support from special bones, and some species possess prominent facial warts. Males tend to be larger, and their tusks and warts are usually more pronounced. Though they have four toes on their feet, pigs walk only on their third and fourth digits, with the remaining two usually held clear of the ground. Their mobile, tasseled tails are effective flyswatters, and the way the tails are held signals the mood of their owners. All pigs have good senses of hearing and smell, and scent-mark extensively.

Wild pigs have been able to adapt to a great variety of habitats and life-styles. The family is widespread in the Old World and has often been successful when introduced into areas outside its natural range. This makes it hard to generalize about pig behavior. Most pigs keep to stable home ranges, but the bearded pig is well known for the lengthy migrations it makes in its native Borneo. Female pigs generally gather with their young into small herds, called sounders, while the males live alone or in bachelor herds, but the pygmy hog is believed to form mixed-sex herds.

Female pigs tend to live peacefully together, but males frequently engage in fights. Although they are not strictly territorial, males will fight over potential mates in the breeding season, and they may fight if they meet at other times.

Adult wild boars and pygmy hogs fight "shoulder-to-shoulder," trying to rake their opponent's neck and shoulders with their tusks. The skin and hair on these parts is thickened as a form of protection, but old boars can still carry deep scars. Younger animals engage in a series of shoulder nudges without tusk blows, so that combat is more a trial of strength than a violent encounter.

In wart hogs and bush pigs, fighting involves the crossing of snouts and tusks in a head-on confrontation. Facial warts help to protect the animals' long snouts by deflecting tusk blows. The giant forest hog also fights head-on, but its tusks are smaller and less important in combat. Instead, it lowers its head and pushes against its opponent using the toughened top part of its skull.

Warts and beards

Of the nine pig species, five belong to the same genus—the pygmy hog, the bearded pig, the Javan warty pig, the Celebes wild pig and the wild boar. The first four of these are confined to parts of India, Southeast Asia and Indonesia.

The pygmy hog lives in savannah grassland, in the Himalayan foothills of Assam in India. It is the smallest of the pigs, weighing up to only 25 lbs., with a body length of 2 ft. The Javan warty pig, found in grassland, swamps and forest on Java and two small islands nearby, is now very low in numbers and may not survive without the help of captive breeding programs. It has three distinctive warts on each side of its snout. Facial warts are also present on the bearded pig, along with a yellowish-white beard over its cheeks. It lives in tropical forests of Malaysia and Indonesia, and is similar in size to the wild boar. The Celebes wild pig of Sulawesi, Indonesia, also lives in tropical forestland and it too possesses facial warts.

Wild boars

Ranging through woodland and grassland over most of Eurasia, the wild boar has evolved into a number of subspecies. It has been introduced by man into North America, New Guinea and Australia, and has a long history of domestication.

PAGE 363 A pair of wild boar and their litter of striped piglets browse a woodland floor. They will eat a variety of plant and animal foods, including roots, leaves, fungi, ferns, insects, frogs, mice and young birds. Boars are the most widespread of the wild pigs, and made up the original breeding stock for the many kinds of domestic pig.

The wild boar is one of the larger members of the pig family, weighing up to 450 lbs. and reaching 6 ft. in body length. It has no warts and only short tusks. Size is one way in which the subspecies tend to differ. Larger animals are generally found in the north of the range, while smaller individuals are often found on islands. The wild boar from the Mediterranean island of Sardinia, for example, may weigh only 130 lbs. Wild boars enjoy a broad diet, consisting mainly of plant foods, but they will also eat earthworms, insects, eggs, nestlings, frogs and mice. In cultivated areas, they dig up roots and tubers of vegetable crops, and can become serious agricultural pests. On the other hand, small numbers of wild boar may be of benefit to woodlands, because they destroy large numbers of pests, such as mice and insect grubs, and can help spread the seeds of woodland trees. To rid themselves of their own pests—bodily parasites—boars frequently take mud baths by wallowing in muddy pools.

Breeding and courtship

Female boars are usually able to breed from the age of 18 months. Males, although mature at this age, may only be able to breed when they are two or more years older and sufficiently well developed to compete for mates with rival males.

In Europe, the breeding season is late autumn, but in moist tropical areas, boars breed throughout the year. Mature males can tell which females are ready to breed by smelling their urine. They then follow them around, emitting low, rhythmic grunts and producing large amounts of saliva. The smell of chemicals in the saliva and the sound of the grunts help to prepare the females for mating. When ready they indicate their willingness to mate by standing still with their backs slightly bowed.

Up to 12 young are born after a gestation period of 115 days. The female gives birth alone, in a snug nest of branches, plants and hay. The piglets are very small and vulnerable at birth, each weighing 18-32 oz.—about one percent of their mother's bulk. They are susceptible to cold, and so for the first 10 days of their life they remain in the nest, huddled together for warmth. Though their parents are a brownish-gray color all over, the young boars have distinctive striped coats.

The piglets are weaned when they are about three months old, but they remain with their mother until the next litter is born. Mother and young may be joined by other families so that small groups, or sounders, are formed. Members of the sounder are usually related, so the group is essentially a clan of

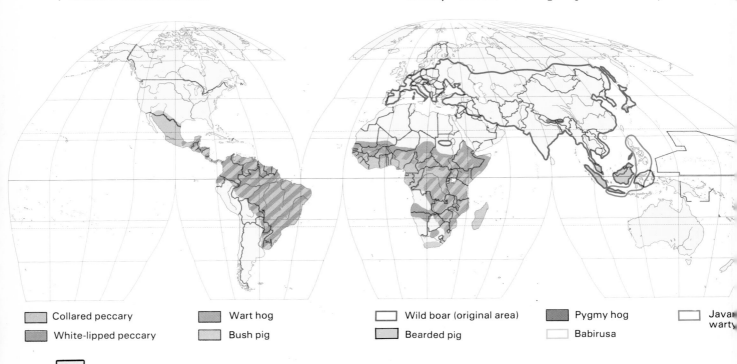

Collared peccary	Wart hog	Wild boar (original area)	Pygmy hog	Java warty
White-lipped peccary	Bush pig	Bearded pig	Babirusa	

females and their offspring. Large sounders of up to 100 animals will eventually split up and occupy adjacent home ranges. Within these ranges the females and young forage for food. Adult males tend to live alone, joining the sounders only during the breeding season.

Domestic pigs

Of all the pig species, only the wild boar has been widely domesticated. The familiar farmyard pigs are all descended from this animal. The earliest domestications probably occurred in Mesopotamia, and records show that there were domesticated pigs in Jericho in 7000 BC. Centuries of breeding have produced over 300 varieties throughout the world, some able to accumulate body weight exceedingly fast. There are many instances of domesticated pigs escaping to the wild and hybridizing with boars. The central Italian form of the wild boar is now almost extinct as a subspecies because it has interbred so extensively with local domestic pigs that have reverted to the wild.

African pigs

Of the remaining four pig species, three live only in Africa. The bush pig, or Red River hog, is found throughout sub-Saharan Africa and on the island of Madagascar in forests, damp woodland and grassland. Though its coat is red or gray, it sometimes has striking white markings on its face and mane. The biggest member of the family is the giant forest hog, which weighs up to 600 lbs. It inhabits tropical forests and the zones where forests and grassland meet in West, Central and East Africa. Perhaps the best known of the African pigs is the wart hog. As its name

LEFT The map shows the world distribution of the pigs and peccaries.
TOP Two male wild boars fight shoulder-to-shoulder, attempting to rake each other's hide with their short tusks.
ABOVE Before giving birth a female wild boar gathers a mound of branches, plants and hay, forming a snug nest for her young.
TOP RIGHT A Javan warty pig displays its bizarre face as it wallows in a mud bath. Three prominent warts on each side of the animal's head accentuate its strange appearance.

THE WART HOG
— WILD PIG OF THE PLAINS —

Wart hogs inhabit open country over most of Africa, south of the Sahara. In appearance and habits they are undoubtedly one of the most striking members of the pig family.

Wart hogs take their name from four unsightly warts that erupt from their faces. These gristly projections are positioned below the eyes and above the snout, and tend to be more pronounced in males. They seem to be protective features, deflecting tusk blows away from the face when males are fighting one another.

Odd features

Along with prominent warts, wart hogs also have two pairs of tusks protruding from their snouts, the upper pair of which are long and sickle-shaped. Their heads are particularly large in relation to their bodies, their skin is sparsely covered in bristles, apart from long dark manes that run along their backs, and the animals' thin, tasseled tails stand erect when they run. With such an odd collection of features, wart hogs have been described as "grotesque" and "prehistoric." But odd though they may look, these animals are remarkably well adapted to life on the African plains.

Wart hogs feed mainly on grasses, nipping off the tender growing tips with their specialized incisor teeth and digging out the roots during the dry season. Contrary to popular belief, wart hogs do not use their tusks to dig for food. Instead, the nasal bones supporting their long snouts are strengthened, making the snouts powerful rooting implements.

To provide shelter on the open plain, wart hogs dig lairs under rocks and thickets. They will also take over abandoned aardvark burrows, providing a secure underground chamber reached by a tunnel. The basic social unit is the family group of mother, father and up to four young. Adult wart hogs are monogamous—staying with one mate rather than breeding with different individuals.

Although the male leaves the female immediately after mating, he returns when the offspring have been born. Birth takes place between October and

November, some 25 weeks after mating. Usually two to four reddish-brown young are born, but litters of up to six are known, even though the wart hog sow has only four teats with which to suckle her offspring. The young remain in the lair for several days before venturing out with their mother onto the open plain.

Courageous defenders

Male wart hogs frequently battle with each other for dominance, fighting head-on and clashing their tusks, but they are also noted for their courage in standing up to other animals. If a young wart hog is attacked by a cheetah out in the open, one of its parents may well launch a furious charge against the predator. The wart hog's speed and its approaching tusks are often enough to scare off the cheetah. In most circumstances, wart hogs retreat to shelter when predators are at large, but

they display a formidable last line of defense against any carnivores that follow them. The adults usually reach the burrows last, turning and backing into them rear-end first. Their retreating action leaves the sharp tusks facing predators, blocking the tunnel entrance, and warding off animals as powerful as lions and leopards.

FAR LEFT A female wart hog and her three young drink from a waterhole. The growing tusks are just visible on the young animals, and their long-haired manes are already well developed.
RIGHT A male wart hog keeps alert for predators as it drinks. Its long snout bears four facial warts that help to protect the animal's face from tusk blows when it is fighting with other aggressive males.
BELOW Six young wart hogs follow their mother, their tasseled tails held aloft. Wart hogs can use their mobile tails as flyswatters.

The natural distribution of the pig family covers most of Europe, Asia, Indonesia and Africa.

suggests, it has four prominent warts on the snout. It also has two sets of tusks projecting from its mouth, thinly haired skin, and a particularly large head. The animal is widespread on open ground over most of Africa south of the Sahara.

The last member of the family, the babirusa, is confined to forests on the Indonesian islands of Sulawesi, Togian, Sulu and Buru. Like the wart hog, the babirusa has two pairs of tusks, but the animal's upper tusks protrude so far through the hide on the snout that they look like a pair of horns. Local people sometimes call this pig the "horned hog."

The peccaries

Peccaries are the American relatives of the pigs, found mainly in the tropical zone from the southwest USA to northern Argentina. Only three species belong to the family: the collared peccary, the white-lipped peccary and the Chacoan peccary. Although pig-like in general appearance, they differ from the pigs in several ways. They are smaller, have a more vegetarian diet, and though their canines are pointed they do not protrude as large tusks. They have complex stomachs and digest plants more completely than pigs.

Peccaries have tough coats made up of thick hairs up to 9 in. long, and they have a large scent gland on the rump. The collared and the white-lipped peccaries share similar habitats—wet and dry tropical forests, chaparral (areas of dense, tangled brushwood) and grassland—and their ranges extend over most of tropical Central and South America. The Chacoan peccary is more limited in range, occupying the thorny forests of the Gran Chaco region of northern Argentina, southeast Bolivia and western Paraguay. Although ancient remains of the Chacoan peccary were found in 1930, the animal was not known to zoologists as a living species until the mid-1970s.

Peccaries are most active in the cooler hours of the day and during the night. They grub around for food on the forest floor, making up for weak eyesight and poor hearing by having a keen sense of smell. They can detect the scent of buried food, such as edible bulbs, located several inches under the surface. The collared peccary is mainly vegetarian, using its snout to root for berries, fruit, seeds and bulbs. But, like the pigs, it will tackle most sources of food including grubs, insects, small vertebrates and even snakes—the animals appear to be immune to the

PIGS AND PECCARIES CLASSIFICATION

The pig family, the Suidae, and the peccary family, the Tayassuidae, both belong to the suborder Suina.

The pigs are divided into five genera, with a total of nine species between them. Of these, five belong to the genus *Sus*. The wild boar, *Sus scrofa*, is the most common and widespread member of the genus, occurring naturally in Europe, Asia, Indonesia, Japan and North Africa. It has also been introduced into North America, New Guinea and Australasia. The pygmy hog, *S. salvanius*, is found in savannah grassland in the Himalayan foothills of Assam in India. The bearded pig, *S. barbatus*, is found in Malaysia, Sumatra and Borneo, the Javan warty pig, *S. verrucosus*, occurs in Java and the nearby islands of Madura and Bawean, and the Celebes wild pig, *S. celebensis*, occurs in Sulawesi.

The babirusa, or "horned hog," *Babyrousa babyrussa*, is another member of the pig family native to Indonesia, living on the islands of Sulawesi, Togian, Sulu and Buru. The remaining three species are found in Africa. The wart hog, *Phacochoerus aethiopicus*, is widely distributed in woodland and grassland south of the Sahara. Its range is similar to that of the bush pig, or Red River hog, *Potamochoerus porcus*, although this species also occupies Madagascar. The giant forest hog, *Hylochoerus meinertzhageni*, occurs in Central, West and East Africa.

The peccaries are the New World relatives of the pigs, and number three species. The collared peccary, *Tayassu tajacu*, ranges from the southwest USA to northern Argentina, and the white-lipped peccary, *Tayassu pecari*, from southern Mexico to northern Argentina. The Chacoan peccary, *Catagonus wagneri*, is found in the Gran Chaco region of northern Argentina, southeast Bolivia and western Paraguay. This species was not discovered until 1975.

venom of rattlesnakes. Both Chacoan and collared peccaries include cacti as an important part of their diet. The white-lipped peccary relies less on vegetable food, taking worms and insects and scavenging on carrion, in addition to fruits and roots.

Social animals

Unlike pigs, peccaries are highly social creatures, living in mixed-sex herds numbering from two to ten animals in the case of the Chacoan peccary, to as many as 100 or more in white-lipped peccaries. Smaller family units make up large herds, and when food is not abundant these family groups may break off to look for food on their own.

Peccary herds occupy stable territories which they scent-mark and defend from intruders. Collared peccaries defend territories of from 75 to 620 acres in area, the size depending on the availability of food and numbers in the herd. Secretions from the animal's rump gland are used to mark trees and boulders within the territory. When the animals move from one site to another, it is usually one of the females that leads the group.

At the beginning of the day, peccaries often spend some time in close physical contact, grooming and scratching each other, and indulging in playful rough-and-tumble. Members of a group also mark each other by smearing secretions from a gland below the eye onto another's cheeks and head. They have a keen sense of smell, and the scent marks aid group recognition, helping to keep the herd together.

Protecting the young

Female collared peccaries can breed at the age of about seven months, while males become sexually mature at about 11 months. Before mating, partners investigate each other's scent glands, but there is no courtship display. Females may mate with several males, but young and subordinate males are prevented from mating by the more dominant members of their sex.

One to four young are born after a gestation period of about five months. They remain dependent on their mother for a further six months, even though weaning occurs at six to eight weeks. In times of danger, both parents and non-parents will shelter young peccaries beneath their bodies. When a herd with young animals is threatened by a mountain lion or a jaguar,

TOP **A babirusa investigates a scent-marked tree. The animal's strange upper tusks grow upward through the snout, curving backward toward the head to resemble a pair of horns. For this reason the babirusa is also known as the "horned hog."**

ABOVE **Two bush pigs forage on a patch of soil, showing the distinctive reddish color of their coats. They use glands on their feet to scent-mark areas in their home range. Bush pigs inhabit forests, woodland and grassland throughout sub-Saharan Africa.**

the herd cannot run away without a high chance of losing a piglet to the predator. In such cases, a young animal may approach the predator and distract its attention while the herd escapes. Such apparently unselfish behavior is found in other animals that live in closely related groups and seems to be linked to the importance of passing on the family's genetic material to future generations.

Peccaries have a wide range of calls, largely connected with warnings or social behavior. Vigilant males give a repeated woofing call to signal alarm and a gentle cough to recall the wandering members of a group. Young animals caught out alone utter shrill clucking calls until they are found. When angry or annoyed, the animals produce a rasping sound by rapidly grinding their teeth together, and when involved in aggressive encounters with one another peccaries may utter laughing calls.

UNDER THREAT

THE PYGMY HOG

The smallest of the wild pigs, the pygmy hog, is now the most seriously threatened member of the pig family. It once lived in grassy swampland over much of the Himalayan foothills from Nepal through Bhutan to northeast India. Today it survives only in the Assam region of India, and in the 1970s the total pygmy hog population was thought to be no more than 150. Since then, pressures on the animals have continued—chiefly through destruction of their habitat resulting from the growth of human settlements and the burning of vegetation during the dry season. The animals presently exist in several forest reserves and the Manas Wildlife Sanctuary in Assam, but in each area their numbers are now extremely low.

Several other wild pigs are now endangered species. The bearded pig is hunted extensively in many parts of its range, and has suffered drastic population declines on some islands. The Javan warty pig is threatened by habitat loss, and by shooting and poisoning on agricultural land, where it is considered a pest. Though protected by law, the babirusa is still hunted in Indonesia, and the giant forest hog of Africa has suffered from the reduction and fragmentation of its natural habitats.

TOP A group of white-lipped peccaries doze in the sunshine. Peccaries are more sociable than wild pigs, living in well-organized mixed-sex herds. Each herd is made up of small family groups that may separate to forage when food is short.

ABOVE A white-lipped peccary shows the long coat, white muzzle and absence of tusks that distinguish peccaries from members of the pig family. Pigs are native to the Old World, but peccaries are New World animals of Central and South America.

HEAVYWEIGHT WALLOWERS

The barrel-shaped hippopotamus—"river horse"
of the ancient Greeks—escapes the blazing
African sun by lying in water.
On land, it is surprisingly agile and fast

The hippopotamuses live in rivers, lakes, swamps and grassland in sub-Saharan Africa.

At first glance, hippos and rhinos may appear similar. In fact they are very different, for whereas the rhinos are odd-toed ungulates related to tapirs and horses, hippos are even-toed ungulates whose closest living relatives are the peccaries and pigs. The earliest remains of recognizable hippos date from the late Miocene epoch, some 10 million years ago. They appeared first in Africa, then spread north into Eurasia about two million years ago. During the Pleistocene period about 100,000 years ago, the hippopotamus was found as far north in Europe as southern England—hippo remains have been found in central London. Of the several species that evolved, only two have survived into the modern era: the hippopotamus, found in and near the rivers, lakes and swamps of much of Africa south of the Sahara, and the pygmy hippopotamus of West Africa.

A shrinking habitat

The hippopotamus was originally found throughout Africa from the Cape of Good Hope to the Nile Delta (with the obvious exception of desert areas), and also in some parts of Israel. In the last 200 years its area of distribution has been greatly reduced by land drainage and hunting. Today it is only found south of the Sahara and north of the Limpopo and Zambezi rivers. The southernmost colony lives in the Addo National Park near Port Elizabeth in South Africa, but this is the result of a recent reintroduction. The largest natural colony in southern Africa is to be found in the St. Lucia Game Reserve on the Indian Ocean coast of northern Natal, South Africa.

The most densely populated area is the lake region of East Africa, or more specifically the Murchison Falls National Park, the Virunga Park in Zaire, the Queen Elizabeth National Park in southwestern Uganda and the Gorongosa National Park in Mozambique. In some places overpopulation has led to significant damage to the environment and has prompted a number of control schemes. Overall, though, hippos appear to be declining both in numbers and range.

The hippopotamus was well known to the ancients, and its name is derived from the ancient Greek for "river horse." Since the animal spends much of the day in the water, there is some merit in this description, but it is hard to see any resemblance to a horse. It has a massive, barrel-shaped body supported by short legs, and a mature bull hippo may be 5 ft. high at the shoulder, 15 ft. long and weigh as much as 10,000 lbs. It has a big head with a huge mouth equipped with up to 40 teeth. The canine teeth are particularly large, and the lower canines of bull hippos develop into razor-sharp tusks up to 20 in. long and weighing more than two pounds each. The prominent eyes, ears and nostrils are all mounted high up on the head and snout, allowing the hippo to wallow in water for hours at a time with only the upper part of its head showing above the surface.

The pygmy hippo

The much smaller pygmy hippo is similar in shape to the hippopotamus, but its legs are longer and the head is smaller in proportion to its body. The eyes are placed on the side of its head rather than on top, making it less adapted to the aquatic life than its larger relative. Indeed, it spends far less time in the water. It lives among the forests and swamps in scattered areas of West Africa, where it feeds on grasses, roots, fruit and foliage. The pygmy hippo is a rare and elusive animal. It was unknown to Western science until its discovery in Liberia in 1841. For several decades it was considered merely a dwarf form of the hippopotamus rather than a species in its own right.

Playing it cool

The main reason why hippos spend so much time in the water is that their smooth, hairless hide has a thin outer layer that loses moisture. Out on the African plain, the hippopotamus would rapidly dehydrate under the hot sun if there was no water to retreat to. Hippos prefer grasses that grow in open, sunny areas. To avoid dehydrating they feed at night and lie in mud or water during the day. The threat of dehydration is of little concern to the pygmy hippos, which feed on leaves in the shady forests.

Water is not only necessary to prevent dehydration, it also keeps the hippos cool—an important consideration for a large animal in a tropical climate. Big animals have a lot of bulk in relation to their surface area and generate more heat than they can lose. The problem is worse for the hippo, which has no sweat glands. Another advantage of wallowing in water all day is that it takes the weight off their legs and conserves energy. Hippos extract relatively little value from their food; lounging in the water all day cuts energy expenditure to a minimum.

ABOVE Water splashes high above two rival male hippos as they battle for domination of a pool. RIGHT The drawings show three stages of a typical fight for supremacy. First one hippo threatens the other with jaws agape. The other responds likewise, displaying his fearsome canines. Finally they lunge at one another mouth-to-mouth.
PAGE 373 Looking from the air like a school of whales, a herd of female hippos and their calves congregate in a Zambian river to escape the heat.

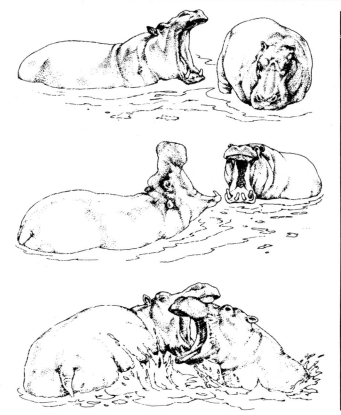

A protective mucus

The hippo's skin has pores that secrete a reddish mucus—a feature that once gave rise to the idea that hippos sweated blood. If the animal stays out of water for some time, this mucus dries to form a coating that protects the skin from the sun's ultraviolet rays and acts as a barrier against sunburn. The hide is very thick, especially along the flanks, where it gives protection against the bites of rival hippos. It is common for males to be heavily scarred, but the wounds heal quickly; this may be because the red mucus acts as a disinfectant destroying any micro-organisms picked up from dung-laden mud and water.

Although hippos spend much of their time in rivers and lakes, they get very little food from the water. A survey of the stomach contents of several animals produced 27 types of plants—and not one of these was

ABOVE A fight between rival male hippos is all that disturbs the placid waters of an African lake. Hippos rest during the day and lounge in the water, conserving energy for the nightly trek to the grazing areas. A hippo does not feed in the water: it simply uses it as a warm bath that prevents its skin from drying out, keeps its body temperature constant, and takes the weight off its feet.

Grazing by night

At sunset, after a day spent lounging in the water or stretched out on nearby mudbanks, the hippos stir themselves and begin to move off toward their pasture grounds. They will travel more than two miles from the nearest water to find their favorite grasses. Isolated pools often act as "staging posts," allowing the hippos to move much further away from their home lake or river, extending their range up to five miles or more.

Hippos eat a lot of grass—an average of 90 lbs. in a night—tearing it up with their lips instead of nibbling it off with their teeth. As a result, many of the plants are torn out by the roots and get no chance to regrow. This explains why hippos are sometimes considered destructive to riverside vegetation. A dense population will destroy the shallow-rooted grasses and leave the sturdier species; this makes the pasture less inviting to both hippos and other grazing animals.

Ideally, the maximum number of hippos grazing in any one area is about 20 animals per 250 acres. But in some areas the density is too high for the pasture available. In the more highly populated parks it is common for two animals to occupy as little as 12 acres, equivalent to 40 per 250 acres. In 1958, it was decided to keep numbers down in the national parks of Uganda by starting a culling program. Within a few years the pasture had improved for all the grazing animals in the park. Unfortunately, political upheavals led to the scheme being abandoned.

Signposts of dung

At dawn the hippos return to the water along traditional paths marked by dung piles. These dung piles are not territorial indications; they simply serve to mark out the route and allow the hippos to sniff their way to the pasture on the darkest nights. The heaps of dung may form deposits up to 2 ft. high. Hippo pathways can resemble cart tracks or even roadways, and often consist of parallel paths separated by a central rise. The hippos travel along them singly or with their calves, and each animal has favorite tracks that lead to preferred grazing areas.

In some cases the hippos show remarkable agility, and the tracks may be carved out of steep, rugged bank sides or forged through dense vegetation. Such tracks can prove tempting to human travelers, but they use them at their peril. When a hippo is alarmed

an aquatic species. Since they do not forage in the water, hippos do not need to be efficient swimmers, and in general, they prefer quite shallow water about three feet deep, where they can walk on the bottom without submerging. They avoid fast-flowing rivers and keep to bends and inlets where the water is calmer. However, they are quite capable of swimming if necessary, helped by their webbed toes, and they can close their nostrils and stay underwater for five minutes at a time.

it is quite likely to set off along its track at a headlong gallop. A three-ton mass of hippo moving at speeds of 20-25 mph is unstoppable, and several unwary travelers have been killed in this way. Although normally harmless, hippos can be aggressive in the water, and males, or females with young, will sometimes attack people in boats if they come too close. An adult may attack without warning and should be approached with care. They may also be dangerous if their escape route into the water is blocked.

Feeding the fish

The hippopotamus plays an important part in African river ecology. It normally deposits its dung in water, shaking its tail around wildly at the same time to scatter the excrement over a wide area. The feces contain a lot of undigested matter that can be reused by other creatures, and certain fishes. Other fishes get their food directly from the hippo itself as it wallows, feeding on the parasites that often infest it.

Parasites on the hippos are also eaten by certain birds, such as the oxpecker. Hippos make useful floating perches, too, for fish-eating birds such as storks, cattle egrets and black rails. The hippos rake up and disturb the riverbed; the turbulence brings new life to stagnant water, breaking up vegetation that may be blocking the river's flow and encouraging the growth of bacteria that deoxygenates the water.

Adult hippos have very few natural enemies. Crocodiles may threaten them, but the reptiles are at more risk from the huge well-armed jaws of the hippopotamus than vice-versa. Cases of lion attacks have been reported on dry land, but the sheer bulk of an adult hippo usually prevents the encounter from going any further than the initial rush. Young hippos are more vulnerable, and if they are left unprotected they may fall easy prey to lions, leopards or crocodiles. To prevent disaster from striking, the mothers rarely leave their young unattended.

The average mortality is about 45 percent during the first year of life, but once they have got through the first year the death rate falls sharply to 15 percent. A two-year-old individual living in an area where it is not hunted by man has every chance of surviving unscathed to the age of 41. This is the average life span in the wild, but a hippo may live longer under perfect conditions—probably up to 45 in the wild and a record 49 years in captivity.

ABOVE Ever attentive, a mother pygmy hippo stands guard over her young. Pygmy hippos are much rarer than the larger species, and more elusive. They live in forests and swamplands, where the local climate is cool and moist, and as a result they spend far less time in the water. This way of life was probably shared by the forest-dwelling ancestors of the larger hippopotamus, but abandoned when they moved to the open plains.

Territory and breeding

Hippos usually live in groups of 10-15 animals, although larger colonies of up to 150 or more animals are quite common. The males are often solitary and territorial; more successful males live with groups of females and young that they defend against rivals. A powerful male may hang onto his territory for eight years or more, but he will have to fight off frequent challenges to his authority. Many territories change ownership every few months.

The males threaten each other by opening their mouths to display the formidable jaw and long canines, grunting, lunging and rearing up in the water, splashing down and scooping water at one another with their vast mouths. If it is intimidated enough, one of the adversaries may back down with a gesture of submission: closing its mouth and turning its body sideways. In theory, the dominant animal's

HIPPOPOTAMUSES CLASSIFICATION

There are two quite distinct species of hippopotamus in the family Hippopotamidae. The most widespread (though its range is now fragmented and reduced) is the hippopotamus, *Hippopotamus amphibius,* found in rivers and grasslands in sub-Saharan Africa as far south as the Zambezi. The much smaller pygmy, hippopotamus, *Choeropsis liberiensis,* is found in isolated areas in West Africa, chiefly among the forests and swamps of Liberia and the Ivory Coast.

aggression is defused by this gesture, but the rules of engagement do not appear to be highly developed among hippos, since many males come out of such encounters with deep gashes in their backs and flanks, inflicted by their rival's sharp canines.

Sometimes, unusually for mammals, the combats end in the death or severe injury of one of the contenders. This may be partly explained by the fact that such battles often take place in overpopulated areas where the hippos have difficulty finding food. With no natural enemies, the animals may be resorting to extreme methods to keep their numbers down.

The birthrate of hippos changes according to the climate. In dry seasons it may drop as low as 6 percent but increases to over 35 percent in wet years. Mating usually takes place in the water, and the female is often totally submerged, although she has to force her way to the surface occasionally to breathe. Most matings occur toward the end of the dry season, so that after a 240-day gestation the young are born during the rains—the most favorable time of year. Typically, each mother has a single calf.

The young hippos may be born on land, but very often the birth takes place in the water. As soon as it is born, the young hippopotamus has to paddle to the surface to take in air; it then goes back underwater to suckle. It does this with its ears and nostrils closed, breaking off at intervals to surface and take a deep breath before returning for more milk. In deep water it is common for the mother hippo to carry her young on her back. She is fiercely protective and will defend the baby vigorously against attacks by predators such as leopards and crocodiles.

The young hippo is weaned at eight months, but it stays with its mother for some time after it has begun to eat solid food. Males become sexually mature at about seven years old, females at nine. Despite this the males remain socially inferior for another ten years at least, and rarely get the opportunity to breed before the age of about 20.

UNDE THRE

THE HIPPOPOTAMUSES

Hippopotamuses were once abundant in rivers over most of the African continent, but remorseless hunting has had a disastrous effect on their numbers. Since historic times they have been entirely wiped out across their northern habitat; in Egypt, the last hippo disappeared in about 1816. In the west and south of their range they are either extinct or reduced to small, isolated populations. For example, in 1987, there were some 650 hippos in the St. Lucia Game Reserve and Park in Natal, South Africa.

Large populations of hippos still exist in national parks in the upper Nile Valley and other areas of East Africa, but hippos are generally in decline and their range is steadily shrinking.

Many are shot for the damage they do to cultivated crops around human settlements, but the majority are hunted now, as in the past, for their large canine tusks (which are harder and finer than elephant tusks), meat and fat, and as trophies. The hippo, nevertheless, has protected status in most African countries.

In West Africa (Liberia, the Ivory Coast and, to a lesser extent, Sierra Leone and Guinea), the pygmy hippopotamus has become increasingly rare. Illegal hunting and destruction of its habitat have reduced the numbers of this small hippo, though there have probably never been many of the species. It is classified in CITES (Convention on International Trade in Endangered Species) list.

ENDURING HARSH LANDS

Hardy and long exploited by man,
camels and llamas are superbly adapted
to life in arid regions, where they can survive
on the barest vegetation

Vicuna

Guanaco

Bactrian camel

Guanaco (immature)

Camels and llamas are ruminants, although their digestive systems are not as complex as deer and cattle—they have three, not four, stomach chambers. They deal with indigestible fibrous food in the same way, swallowing large amounts at a time and working on it at leisure by chewing the cud. Their digestive systems are specialized in other ways, and desert-living camels can survive on dry, thorny vegetation that would not support other mammals. Camels and llamas stand on well-developed pads (formed by the sideways expansion of the foot bones) instead of hooves.

The origin of the camel family can be traced back to the Eocene epoch, between 40 and 60 million years ago. Like the horses, they first appeared in North America, evolving slowly over tens of millions of years. About three million years ago, primitive members of the family began to make their way into Asia and South America, where they evolved separately into camels and llamas.

On the steppes of Asia

The Bactrian or two-humped camel is a central Asian species that was once found over much of the continent, in Turkey, Iraq, Iran, Afghanistan, India, Mongolia and China. It takes its name from the region of Bactria, north of Afghanistan, that was once thought to be its original home. In many of these areas the winter temperatures drop well below freezing, and the Bactrian camel grows a long, thick winter coat that gives it a shaggy, clumsy appearance. In spring, when the long coat is molted, the Bactrian camel looks even more unkempt, although it retains a thin mane on its chin, forequarters and humps.

The Bactrian camel was probably first domesticated over 4000 years ago, and many animals are still kept for use as pack animals. They tend to be more heavily built than the wild Bactrians of the Gobi Desert, with bigger humps that are often folded over to one side.

ABOVE RIGHT In spite of the Bactrian camel's improbable appearance— lopsided humps, leggy build and shaggy, moth-eaten coat—the animal is well adapted to its harsh habitat in central Asia. PAGE 379 Dromedaries gather around a waterhole.

A dromedary may go for up to ten months without drinking, but it must take in moisture by feeding on succulent desert plants. PAGES 382-383 Outlined by the evening light, a family of vicunas pick their way across the sparse mountain pasture.

Ships of the desert

The dromedary, also called the one-humped or Arabian camel, is much the same size as the Bactrian camel (both species weigh 1000-1500 lbs. when fully grown) although it tends to be slightly taller, measuring up to 7 ft. 6 in. at the hump. It is very similar in build, apart from having one hump.

Dromedaries are superbly adapted to life in the desert. They have long eyelashes, nostrils that can be closed to keep out the sand, and broad foot pads that enable them to walk on soft sand without sinking in. Most important, however, is their ability to endure long periods in the burning heat of the desert without drinking. If they are not working, they can survive for up to 10 months without water, enabling them to forage far out in the desert, well away from other grazing animals. This gives them a monopoly on the desert vegetation, although few other animals would care to eat the dry, prickly plants that make up much of the dromedary's diet even if they had the option.

At one time it was thought that the dromedary stored water in its hump, releasing it drop by drop over the months to the rest of its body. This theory has now been discarded. The camel does not store water; rather it conserves it very well. One way it does this is by producing extremely dry feces and highly concentrated urine containing little water.

THE VICUNA
— FAMILY LIFE IN THE HIGH ANDES —

The social organization of the vicuna is based on small family groups. A typical family is made up of an adult male, a number of females with their offspring and a few immature, year-old females. These groups rarely exceed ten in number and are often much smaller. The male is the head of the family. He guides the group to the pasture grounds and back to a chosen sleeping area to rest at night, watches over them as they graze, and wards off intruders. Vicunas are very jealous of their territories, and the males will vigorously defend both the resting and grazing grounds, which are often some distance apart. They are linked by a "corridor" that is not part of the territory and is not defended.

The grazing territories are usually areas of pasture some 45 acres in size, situated on fairly fertile, flat or rolling ground with a perennial water supply. The resting territories are much smaller, averaging only 6 ½ acres, staked out on higher land among rocks, and if possible protected from the wind and prying eyes by a ridge.

Bachelor herds

In addition to the family groups, there are also groups of immature males. Numbers vary from two to 100, but in general such groups rarely contain more than 25 animals. Such a herd does not have a well-defined structure and is merely an assembly of aimless individuals. The bachelor herd does not have defined territories; instead the animals wander around at random and often trespass on the grazing lands of other vicunas. They are always attacked and chased off by territorial males.

Mature males are solitary animals until they gather their families together. When a male vicuna feels ready to strike out on his own, he leaves the bachelor herd and attempts to win his own territory and start a family. This does not happen all at once, however. The male starts by wandering alone for some time, checking out the options; if the population density is high, it may be some time before he finds a vacant area that he can call his own and can begin to collect his harem.

Each family group is self-regulating, in that the young vicunas are ejected from the group as soon as they are able to lead an independent life. This means that each area is grazed by the same number of animals year after year, an effective way of preventing the overgrazing of sometimes scanty mountain pasture.

The rejection process begins early. Adult males start to show hostility to their male offspring when the young are only three months old, but their aggression is countered by the protective reaction of the mother.

As the young male grows, however, the mother starts to lose interest in defending it. The adult male becomes more and more aggressive until eventually the protective instinct of the female is overcome and the young animal is expelled from the herd.

Young females are expelled in a similar fashion. Shortly before the mating season in spring, both the male and the adult females attack immature females with increasing frequency until they are forced to leave. This apparently heartless system has its advantages, however. It ensures that the animals are well distributed over the available land, and it also means that there are plenty of females available for young males setting up on their own.

LEFT A group of vicunas raise their long necks to check the horizon as they pick their way down to a watercourse for a drink. The permanent territory of a vicuna group often centers on a source of water.
TOP RIGHT A young vicuna takes time out to chew the cud. The thick, soft pelt of the vicuna has always been highly valued, and until the 1970s the survival of the species was threatened by hunting. They are now fully protected, and their numbers are increasing. The current population stands at over 80,000.
RIGHT An immature vicuna is chased from the grazing territory by an adult.

The camel survives over the hottest months because it is able to withstand levels of dehydration that would kill other mammals. Man, for example, can survive a maximum water loss of 20 percent of his body weight in a cold climate. In a hot climate this figure drops to 12 percent, though he starts to lose his faculties at even 5 percent water loss. The water the body loses comes from the tissues and blood, which becomes too thick and slow to circulate properly. In camels, water is not lost from the blood, which continues to circulate normally. As a result, camels can function perfectly even if they lose water amounting to almost 30 percent of their body weight.

Quenching its thirst

When a dromedary gets the chance to drink, it shows just what an awesome capacity it has for water, swallowing up to 35 gallons at a rate of 4 gallons a minute. It can regain its normal weight with one drink, provided this has not dropped by more than 20 percent. A very dehydrated camel requires two to three drinks; even so, it is back in top form in a few hours.

A dromedary can also cope with extremes of temperature better than most mammals because it can allow its body temperature to rise as much as 11°F without apparent ill-effect. Its body temperature can vary from 93°F at night to 104°F during the hottest hours of the day. This means that it does not have to lose heat by sweating until its internal temperature threatens to exceed 104°F; once again, this conserves precious moisture. The hump (or humps in the case of the Bactrian camel) are composed largely of fat and act as energy stores. They allow the animal to go for long periods without eating—a valuable ability in the deserts, where grazing is always sparse.

Most dromedaries live under conditions of semi-domestication. But for four to five months of the year, they are left to their own devices. The meager desert plants and bushes are quite sufficient to keep them alive during this period and during the coldest months a camel can walk for up to 650 miles without drinking.

Rival male dromedaries will often fight to establish dominance. They face one another at a distance, raise their heads and display their manes. As they come closer they gurgle, spit, defecate and beat their tails against their backs and undersides in an attempt to intimidate one another. To fight, they stand neck-to-neck while each tries to push over and bite its adversary.

TOP Unsteady on its legs, a young guanaco takes a look around while its mother keeps watch. Guanacos are the most widespread of the wild llamas, and though protected in Peru and Chile, they suffer badly from hunting in Argentina and are gradually decreasing in numbers. ABOVE With their long wool and hardy constitutions, alpacas make good substitutes for sheep on the high-altitude pastures of the Andes. They are also used for meat and as pack animals.

ABOVE **The multicolored, multipurpose llamas have been domesticated for centuries in the highlands** of South America. They probably descended from the guanaco, but the modern llama is bigger.

Domestication of camels

Both camel species are used as pack or riding animals and are kept for their wool (camel hair), milk, hides and meat. The Bactrian is valued especially for its hair, which is longer and finer than that of the dromedary. Modern domestic Bactrian camels in China, Mongolia and Afghanistan, bred for their hair, have been developed by crossing ancient Bactrian stock with dromedaries, producing an animal with a coat that is resistant to both heat and cold.

Dromedaries can carry over 200 lbs. on their backs and are used chiefly by North African nomads. The camels are important socially, being required as marriage gifts and as repayment for crime. These camels produce as much as 1.5 gallons of milk a day for up to 18 months, and often form the main food for nomadic people in the western Sahara Desert. Mongolians are fond of a fermented yogurt-type drink made from the milk of Bactrian camels, and they also make a camel-milk cheese. Bactrian dung is used for fuel; one camel can produce 600 pieces of dung a night, equivalent to 500 lbs. a year.

South American relatives

The llamas and their relatives are basically small camels without humps. There are four species: the llama and alpaca, which exist only as domestic animals, and the guanaco and vicuna, which are wild. The llama may have been derived from the guanaco but is larger and stronger, possibly because over the centuries it has been bred as a beast of burden and a source of meat. It was domesticated over 4000 years ago, and no wild llamas survive today. The role of the llama as a baggage animal is being undermined by mechanization, but the market for its wool is expanding.

The alpaca was once thought to be a domestic version of the vicuna, but it is now considered to be the result of cross-breeding the vicuna and the guanaco. It is reared mainly for its fine wool, and today it is becoming more popular than the llama, which has a relatively coarse pelt.

The graceful guanaco

The guanaco is about 4 ft. high at the shoulders and weighs up to 175 lbs. It is a slim, graceful animal with a long neck and a thick coat, red-brown on the back and white beneath. Originally it was found from the Peruvian Andes in the north, through Patagonia in Argentina and south to Tierra del Fuego; but despite

Camels range from North Africa through southwest Asia to Mongolia; llamas are found in the Andes.

protection in Chile, Peru and Argentina, it has already disappeared from many places where it was formerly common. Today it is found mainly in mountain areas up to a height of 13,000 ft., although the herds often descend to the lowland plains during the dry season.

The guanaco lives in a variety of habitats from desert grassland to forest and mountain pasture, and feeds on a range of plant material, including both grass and tree foliage. About 95 percent of the guanaco population live in Argentina, where they are still vulnerable to poaching. They also suffer from the erosion of their habitat as the pampas grasslands are developed for agriculture and stock rearing.

Guanacos are very agile animals; the pads on their feet are narrower and more movable than those of the camels, giving the animals an excellent grip on the bare rock of the mountains, and they have a goat-like ability to pick their way across steep, rocky ground. They live in small, harem-based groups consisting of a male, several females and their young. Immature or subordinate males live in separate herds, moving in to challenge the dominant males during the mating season. They engage in violent battles, in which each animal kicks and tries to bite the other's front legs. This involves a continual interweaving of necks and progressive side pushing—one of the most common ways ungulates fight, but one which, in many species, has become ritualized with no attempts made at biting.

Baby of the family

The vicuna is much slimmer and more delicately built than others in the llama group and weighs about 100 lbs. It lives in the high Andes of Peru, Bolivia, Chile and Argentina, at heights of 12,000-16,500 ft., although at one time it was probably found at lower levels. Vicunas are well adapted to arid, dry climates, and the alpine *puna* grassland provides them with sufficient food. They feed mainly on succulent herbaceous plants and spurges (flowering plants). Water is an important element of the vicuna's diet—unlike the guanaco, which gets all the moisture it needs from its food—and during the dry season vicunas will visit streams or springs up to twice a day. They live in mixed family groups and bachelor groups, and each group has two territories: a feeding territory on the grasslands and a sleeping territory at higher altitude, where the animals are less vulnerable to attack.

The vicuna has recently made a dramatic recovery from near-extinction. Numbering an estimated million and a half at the end of the Inca period in the early 16th century, vicunas had fallen to some 400,000 by the 1940s. By 1967 the population was down to less than 15,000 individuals, and it was placed on the list of endangered species. Now accorded full protection, its numbers have increased to about 85,000 and although it is still vulnerable to changes in its habitat, it is no longer in imminent danger.

CAMELS AND LLAMAS CLASSIFICATION

The camel family (Camelidae) consists of the camels of Africa and Asia, and the South American llamas and their relatives.

There are two species of camel. The dromedary, one-humped or Arabian camel, *Camelus dromedarius*, is a widespread, largely domesticated species found in the arid regions of North Africa and southwest Asia; there is also a flourishing feral population in the desert heartland of Australia, descended from domestic animals introduced in the 19th century. The much rarer Bactrian or two-humped camel, *Camelus bactrianus*, has a scattered distribution from Iran to China, the largest wild populations living on the steppe grasslands of Mongolia.

Domesticated Bactrian camels are found in China, Mongolia and Afghanistan.

There are four species of New World llamas, grouped into two genera. The most common are the domesticated species: the llama, *Lama glama*, and the alpaca, *Lama pacos*. Both are found along the Andes, but the alpaca rarely occurs south of Bolivia. The guanaco, *Lama guanicoe*, is the most widespread of the wild species and is found in the foothills of the Andes from Peru to Patagonia. Up to four subspecies have been described. The much rarer vicuna, *Vicugna vicugna*, is a native of the high Andes from central Peru to northern Argentina. There are two subspecies: the Peruvian vicuna and the Argentinian vicuna.

THE SHY WOODLANDERS

Members of the three deer families range from
the rabbit-sized chevrotains and musk deer
to the noble, antlered elks and red deer

Water deer

Musk deer

Lesser mouse deer

Indian muntjac

Southern pudu

Pampas deer

Among the even-toed ungulates, three families are referred to as deer—the chevrotains, or mouse deer; the musk deer; and the true deer, the cervids. The cervids are the deer familiar to most of us, and they are by far the most numerous and widespread family. Though similar in body shape to the true deer, the chevrotains and musk deer form quite distinct groups, and in fact only the true deer possess antlers.

The timid chevrotains

Chevrotains are small, shy creatures, and their alternative name, mouse deer, reflects their timid nature. Unlike most mammals, the males are generally smaller than the females. Fossil evidence suggests that chevrotains have changed little over the last 30 million years, and in both appearance and habits they form a link between the true deer and less specialized ungulates such as pigs. Like the pigs, they have short legs, all four toes are fully developed on each foot, and communication is by scent and voice rather than by visual displays. They do not have antlers and, in the case of the males, long upper canines protrude from the mouth to form small tusks. Like deer, however, they are ruminants—they have a set of complex, multichambered stomachs, and they chew the cud.

There are four species: the water chevrotain is found in the rain forests of equatorial Africa, while the spotted mouse deer, the lesser mouse deer and the larger mouse deer live in the tropical forests of India and Southeast Asia. They forage by night, combing the forest floor for fallen fruit and browsing on low-growing foliage. A clue to their presence is the tunnel-like trails they make through the forest undergrowth. They are rather clumsy creatures, poorly equipped for running, and they depend on the darkness and dense undergrowth to conceal them from predators. The strategy is not wholly successful, and chevrotains frequently fall victim to leopards, large birds of prey, snakes and crocodiles.

The water chevrotain of Africa is the best-documented member of this elusive family. It is a solitary creature, except during the mating season, and usually lives near rivers. It is a good swimmer and will often take to the water when alarmed. Each animal has a home range defined by scent-markers, but there are few territorial conflicts and the chevrotains take little

TOP Little larger than a rabbit, the lesser mouse deer spends most of its life skulking in the undergrowth of the Asian tropical rain forests.
ABOVE A startled water chevrotain looks for an escape route. The largest of the chevrotains (up to 30 in. long), and the only one that lives in Africa, it is an able swimmer and will often dive into rivers to escape predators.
PAGE 389 Red deer stags show off their fine antlers, still covered by the velvety skin that protects them as they grow.

notice of one another for most of the year. During the breeding season the males will fight to establish mating rights, but there is no permanent hierarchy. They use their sharp canine teeth in combat and the specially thickened hide on their rumps minimizes the damage inflicted during these encounters. Some eight months after mating, a single calf is born; it is then concealed in the undergrowth until it is weaned at the age of about three months.

ABOVE A male musk deer shows off his weaponry — a pair of gleaming ivory tusks that can grow up to three inches long. During the mating season rival males lacerate each other with their tusks, and fatal wounds are common. As if this were not enough, the deer are also persecuted by man for the precious musk secreted by the males, which may be literally worth its weight in gold. The substance is used in fine perfumes and Oriental medicines.

Musk deer

The three species of musk deer are found in eastern Asia and the Himalayas. Like the chevrotains, they possess no antlers, but the males have long upper canines. They, too, are fairly small animals, with an average weight of about 25 lbs. Yet they also bear a strong resemblance to the true deer. They have long, graceful necks, small heads with large ears, and long, slender legs.

The hind legs of the musk deer are longer than their forelegs, giving the animals a slightly humpbacked appearance. The tusk-like canine teeth of the male are particularly striking, protruding several inches below the mouth. During the battles that occur in the mating season, the males use these teeth to attack each other, inflicting deep, sometimes even fatal, wounds.

The most unusual feature of the musk deer is the musk bag possessed by the male—a swelling about the size of an apple on the animal's belly. This contains the oily, strongly scented secretions produced by the male's musk gland during the mating season. This secretion is probably a chemical signal to female animals—its precise function is unknown—but it is a substance that has been prized by man for centuries. Dried musk forms the principal ingredient of many perfumes and of traditional medicines in the Far East.

CHEVROTAINS CLASSIFICATION

The chevrotains or mouse deer of the family Tragulidae are a primitive group of small tropical forest-dwelling animals that are intermediate in form between pigs and true deer. There are four species grouped into two genera.

Three of these are found in India and Southeast Asia. The smallest is the lesser mouse deer, or lesser Malay chevrotain, *Tragulus javanicus*, which lives in the Southeast Asian rain forests and mangrove thickets, and is little larger than a rabbit. It shares this habitat with the larger mouse deer, or greater Malay chevrotain, *T. napu*. The spotted mouse deer, or Indian spotted chevrotain, *T. meminna*, lives in the rain forests of India and Sri Lanka, mainly in rocky areas.

The water chevrotain, *Hyemoschus aquaticus*, which is about the size of a hare, lives in the rain forests of Central and West Africa, usually near water.

Musk still provides a flourishing export industry in China. Japan, one of the chief importers, takes up to 11,000 lbs. of the substance each year. Since one musk bag contains about 1 oz., this represents the production of about 180,000 male deer. Today musk deer are farmed and the musk can be gathered without harming the animals—making the wholesale slaughter of wild musk deer unnecessary. Despite this, hunting continues and, along with deforestation, has caused the musk deer to become rare in many parts of its range.

Accomplished climbers

Musk deer live in rocky, mountainous areas covered with dense vegetation and bushes, and feed on a variety of grasses, leaves, roots and shoots. In northern areas they eat mosses and lichens during the winter. They are true ruminants, like deer, and spend several hours of the day chewing the cud. They are accomplished climbers and will even clamber up leaning trees to obtain food, their long lateral toes giving them a good grip on the bark. If alarmed, they take off in a series of bounds and take refuge among rocks or thick vegetation.

For most of the year musk deer are solitary animals that keep within well-defined territories, but in early winter the males enter a phase of sexual activity that

drives them to seek out and chase females, and fight off rivals. They eat little during this period, but expend a lot of energy; and after the mating season is finished they may take some time to recover their condition.

The females give birth to single calves (or occasionally twins) and hide them among rocks and undergrowth, visiting them at intervals to allow them to suckle. After four weeks they are strong enough to follow their mothers, but they are still vulnerable to predators such as foxes, wolves, lynxes and eagles.

Disposable antlers

The true deer, the cervids, are a widespread and successful group of grazing animals. They are found throughout Europe, Asia, North America and South America, and many species have now been introduced in Australia and New Zealand. The 36 species

TOP LEFT An unreceptive female muntjac tosses her head in the air and turns her neck to reject a male's advances.
BELOW LEFT A young male adopts a submissive posture before an adult male.

ABOVE The antlers of an adult muntjac are small and simple compared with those of other deer. Each antler has just one short side branch and is set on a bony projection from the animal's skull.

in the family range in size from the dog-like southern pudu, weighing less than 18 lbs., to the mighty elk, which may grow to 1750 lbs. Though deer are typically woodland animals, their habitats include open grassland in Argentina, tropical forest and Arctic tundra.

Like cattle and sheep, deer are ruminants: their digestive systems are divided into a number of separate compartments containing microorganisms that ferment and break down fibrous vegetation, such as grass and leaves. The food is gathered and

LEFT **A fallow deer stag has distinctive palm-like antlers. Though fallow deer are typically brown in color, with white summer spots, their coat is highly variable, and white-and-black forms frequently occur. This individual is pale over most of its body.**
RIGHT **A female fallow deer takes up an alarm posture. Other alarm signals are barks or bleats and a flash from the black-and-white tail area.**

impressive, multi-branched structures. Unlike the horns of other ungulates, antlers are discarded each year and replaced by new, larger antlers that grow to take their place. During this period they are formed of living, growing tissue, supplied with blood through a network of vessels covered with a soft, smooth skin known as velvet. Eventually the tissue solidifies; the velvet is scraped off and the antlers become completely mineralized dead matter. After a few months in this state, they fall off and the process starts again.

In all but two species of deer, it is the males that possess antlers: in the reindeer, females have them too, while in the water deer, neither sex has them. Their main function is as an aid to sexual display and in asserting dominance within the herd.

Muntjacs

Though muntjacs belong to the cervid family, they possess some characteristics similar to those of the musk deer. They are small deer with short legs and a noticeably humpbacked profile. Like musk deer, the males have large upper canine "tusks," but they also have small, simple antlers mounted on long, bony pedicels extending up from the skull. When the antlers are shed, these long pedicels remain, giving the muntjac a horned appearance throughout the year.

The five muntjac species range over China, Southeast Asia and Indonesia, preferring dense scrubland and forest habitats, where they browse on leaves, fruit, bark and fungi. Like many other deer, they have a good sense of smell and use a range of secretions from their scent glands to communicate with each other. Scent cues are important in a habitat where leaves and bushes limit visibility, especially at night.

swallowed rapidly, then regurgitated later to be thoroughly chewed (chewing the cud). Once chewed it is swallowed again for further digestion. The whole process takes about 48 hours, but it does ensure that most of the food value is absorbed.

The most spectacular feature of deer is their antlers. In some species these are simple spikes, but in others such as the elk and red deer, they grow into

ABOVE Adult chital have reddish-brown coats with numerous white spots arranged in lines over their backs. Their antlers curve up and outward to form a distinctive lyre shape.

Like all deer, muntjacs have scent glands located on their heads: two form a characteristic V shape on their foreheads, and two are found below the eyes, in pits almost as large as the eye sockets themselves. Studies on captive muntjac herds show that the animals will sometimes mark territories using their scent glands; but at other times there may be no evidence of territoriality. When marking does occur, the animals stand upright and rub their heads and muzzles against trees, bushes or the ground. Deposits of dung and urine containing scented secretions may also serve as territorial markers.

Some of the muntjacs have been successfully introduced from Asia into Europe. Reeve's muntjac, also known as the "barking deer," was first introduced in England on the Duke of Bedford's Woburn Estate at around the turn of the century. It has since spread into other parts of southern England, where populations have become established in the wild. Most of the studies of muntjac have been conducted on these introduced animals rather than on those living in their native lands.

MUSK DEER CLASSIFICATION

The family Moschidae consists of three species of musk deer: *Moschus moschiferus, M. chrysogaster* and *M. berezovskii*. They are so similar in appearance and behavior that until recently they were regarded as a single species. They are natives of eastern Asia, from the Himalayas through China to Siberia.

Following the scent

European muntjac follow a seasonal breeding pattern, with mating taking place in early spring. Males find females that are ready to mate by detecting special scent cues in their urine. The males then approach the females with outstretched necks and slightly raised heads, and they may try to lick the females' hindquarters. Licking is an integral part of courtship, helping mates to exchange scent signals. Successful mating is followed by a gestation period of about seven months, after which single young are born. The fawns are weaned after another seven to eight weeks and then leave their mothers.

TRUE DEER CLASSIFICATION: 1

The true deer family, or Cervidae, are a highly successful and diverse group of ruminants (animals that chew the cud). They range over Europe, Asia, North America and South America, and several species have been widely introduced into other parts of the world. In all members of the family except the water deer, the males possess antlers. There are 36 species in 16 genera, and these are grouped into four subfamilies: the Hydropotinae, the Muntiacinae, the Cervinae and the Odocoilinae.

The subfamily Hydropotinae has only one species: the water deer, or Chinese water deer, *Hydropotes inermis*, which lives in swamps in China and Korea. The subfamily Muntiacinae consists of the tufted deer, *Elaphodus cephalophus*, and five species of muntjac from the genus *Muntiacus*. The subfamily Cervinae includes some of the most familiar of the large-antlered deer. There are four genera with a total of 14 species, including the red deer, *Cervus elaphus*, the sika, *Cervus nippon*, and the fallow deer, *Dama dama* (see True Deer Classification: 2). The largest subfamily is the Odocoilinae, with nine genera and 15 species. These include the roe deer, *Capreolus capreolus*, the reindeer, *Rangifer tarandus*, and the elk, *Alces alces* (see True Deer Classification: 3).

Muntiacinae

Two genera containing a total of six species make up the deer subfamily Muntiacinae. They inhabit China and Southeast Asia, and some species have been introduced into Europe.

The genus *Elaphodus* contains only one species, the tufted deer, *E. cephalophus*, found in central, southern and southeastern China and northeastern Burma. The simple, unbranched antlers are partly or completely hidden by tufts of hair which grow on the forehead.

The genus *Muntiacus* comprises five species. The Indian muntjac, *M. muntjac*, occurs in the wild in India, Sri Lanka, southwestern China, Southeast Asia and Indonesia. It has about 15 subspecies. Reeve's muntjac, *M. reevesi*, also known as the Chinese muntjac or barking deer, occurs naturally in eastern China and Taiwan. It has been introduced into southern England, and wild populations have become established in several areas. The black muntjac, or hairy-fronted muntjac, *M. crinifrons*, lives in eastern China. Fea's muntjac, *M. feae*, has a very restricted range in southern Burma and western Thailand. Roosevelt's muntjac, *M. rooseveltorum*, occurs only in Laos in Southeast Asia.

One of the reasons for the muntjacs' success in colonizing southern England is that they are capable of giving birth to a fawn every seven months, if conditions are good. Unlike native British deer, the English muntjac have no fixed breeding season, so fawns may be born any time of the year.

Though muntjacs may live together in groups of up to five within captive populations, they probably lead solitary lives in the wild, except when mating or with young. Captive herds have given naturalists the chance to observe the rivalry that occurs between males during the breeding season. When competing for females, males make remarkably loud barks, repeated over long periods at intervals of about five seconds. Unlike other deer, however, they do not use their antlers in threat or conflict, but use their canine teeth instead.

The adaptable fallow deer

The fallow deer is a remarkably adaptable species—a factor that has contributed to its successful colonization of new habitats around the world. Originally found in the Mediterranean region and the Middle East, it has been introduced to northern Europe and Australasia. Fallow deer usually inhabit woodlands, spending most of their time grazing on the grasses and small plants found in open glades. Yet they can cope with other less favorable environments, and even tolerate snow for a few months. They have been successfully introduced to parklands and are easily tamed.

Fallow deer are medium-sized deer: the males, which are rather larger than the females, stand up to 3 ft. at the shoulder and weigh up to 220 lbs. Their coat color is highly variable; though the normal

summer coloring is fawny brown with white spots on the back and flanks, some animals may be albino (all white) or melanistic (dark). The spots are lost in winter, and the normal coat becomes a more grayish-brown. The antlers of fallow deer are branched, with the angles filled in between the branches to form palm-like shapes. Like other deer, they have a seasonal cycle of antler development. Males grow antlers up to 20 in. in length through the summer. They use them in dominance displays during the autumn breeding season, or rut, and then cast them in winter.

Male and female fallow deer live apart other than during the rut. Females (does) and young live in groups of as many as 70 animals, and the males (bucks) often form bachelor groups. Some areas of woodland are used as "buck areas" and "family areas" at different times of the year. Separation of the sexes probably avoids a situation in which the animals compete directly with each other for food. If all the animals foraged together, the family groups might not be able to find enough food for their needs. Antlered males could readily assert their dominance over females and young to claim the best grazing.

TOP Swamp deer in their natural habitat of damp grassland. Large-scale drainage programs are a serious threat to their population.

ABOVE The heavy body of the hog deer gives this animal its name. It is usually solitary but may be seen in groups of up to five animals. The hog deer is found in marshes and near rivers in northern India, Sri Lanka, Burma, Thailand and Vietnam.
FAR LEFT When antlers have completed their yearly phase of growth, deer may rub them against trees to scrape off the velvet — an activity known as fraying.
LEFT A red deer stag marks his territory using secretions from a gland near the eye.

Fallow deer mate in late autumn, and after 230-240 days the females give birth, usually to a single fawn; twins are very rare. Young females remain with the family groups until their second year, whereas yearling males usually leave before then. Maturity is reached in both sexes by about 16 months, and the full life span may be up to 15 years.

Large numbers of fallow deer can seriously affect the regeneration of woodlands. This is because the deer eat young saplings and prevent new growth. Adult males may also damage trees by thrashing their antlers against them to mark out territories, or by using them as "fraying stocks" to rub the velvet from their horns. For these reasons, it is often necessary to control the size of wild fallow deer populations.

The rediscovered Persian

A type of fallow deer, larger and more red in color than many animals alive today, was once found in Iran, Iraq, Syria and probably in parts of North Africa. The Persian fallow deer is regarded by some zoologists as a separate species and by others merely as a subspecies of the European fallow deer. The Persian deer was a favorite quarry of ancient Mediterranean people, depicted in hunting scenes from that time. Early in the 20th century, it was thought to be extinct.

Then, in 1955, this "extinct" deer was sighted in swampy forests along the banks of the Dez and Karkeh rivers on the Iran-Iraq border. Less than 50 remained, and though sectors of these forests have been set aside as reserves for the deer, the wild population is still very small. A captive breeding population was established at Dasht-e-Naz, north of Tehran, and by 1974 contained more than 30 animals. Other captive herds in zoos and parks may help to save this deer from true extinction in the wild.

The prolific chital

The chital, axis, or Indian spotted deer is another deer species that has proved to be highly adaptable. Native to India and Sri Lanka, it has been able to colonize parts of South America, Europe, Australasia and even the Hawaiian Islands. It is still the most abundant species of wild deer in India, though its numbers have declined in recent years as tree felling and cultivation destroy its habitats.

Chital are similar in size and coat color to fallow deer, with white spots over their backs, but they have distinctive antlers. These curve outward in a lyre shape when seen head-on. Each graceful curve has three branches, and can measure over a yard in length. Chital do not have a seasonal antler cycle in

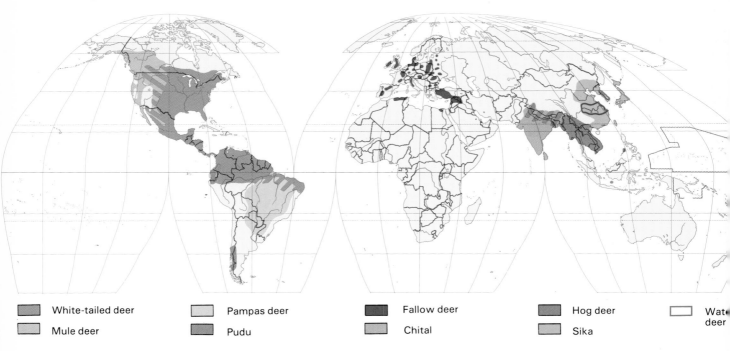

White-tailed deer

Mule deer

Pampas deer

Pudu

Fallow deer

Chital

Hog deer

Sika

Water deer

their native lands, but grow and shed their antlers throughout the year. However, antlers do seem to be cast in a regular pattern among deer that live in a particular locality.

Together with the lack of a seasonal antler cycle, there appears to be no seasonal restriction on breeding activity for most chital. Females may bear young twice in one year. Such a breeding rate enables chital population numbers to rise quickly and reach high densities. However, a seasonal breeding cycle is found in some areas, as in the Khana National Park in India. There, rutting occurs between March and June, and fawns are born from October to December after a gestation period of about seven months.

Mixed herds

Generally, chital live in herds of five to ten animals, but groups of as many as 100 or more may assemble during the monsoon rains. Mixed herds occur, as well as herds composed of either females and young, or bucks. The animals feed almost exclusively on grass, usually at the edge of woodlands.

Although male chital do not keep harems of females, they do establish a type of hierarchy among themselves, using a range of display signals. Dominance is asserted by an exaggerated, stiff-legged walk, with the head and neck raised to show the white throat patch. Another display involves arching the neck and pointing the muzzle downward, while moving the ears up and down, hunching the body and raising the hair on the flanks. In other cases, the head and antlers are lowered as one male walks toward or charges another—a close relationship seems to exist between antler length and hierarchical status.

Unusual companions

Chital are sociable animals and are often seen with other ungulates such as swamp deer, reedbuck or gaur. A more unusual association is that between chital and langur monkeys. Outside the monsoon season, chital have been observed to spend up to eight hours a day browsing in woodlands under the trees where langurs are feeding. The chital exploit the litter of foliage dropped or dislodged from the trees by the monkeys. When the grass is sparse, between November and June, this additional source of food is of great value to the chital and they actively seek out langur troops.

LEFT The map shows the world distribution of some of the true deer.
TOP Eld's deer are distinctively marked with white hair around the chin, the eyes and the ears. They are seriously endangered over many parts of their range.

ABOVE A red deer stag roaring during the autumn rut to ward off rivals. The repeated bellowing sounds may be heard nearly two miles away. If roaring fails to settle rivalry, the stags may fight by locking their antlers together and engaging in a pushing match.

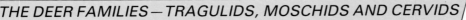

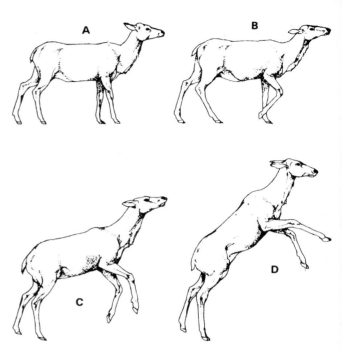

ABOVE A female red deer emerges from a river followed by her fawn. For many months of the year, red deer herds are made up only of females and their young; the males join the herds during the mating season. ABOVE RIGHT Aggressive behavior of a female wapiti: the deer lowers its ears slightly (A); leans forward (B); stands on its hind legs (C); and paws the air (D).

Since chital and langurs respond to each other's alarm calls, they both benefit from feeding together by being better able to detect the approach of predators, such as leopards and tigers. Although langurs sometimes descend from the trees, they tend to remain aloft. This means that the monkeys will be able to detect predators from the trees, while chital are able to locate predators from the ground. Since langurs have good eyesight, and chital a good sense of smell, they can also pool their senses together to guard against danger. Although both animals seem to benefit from the association, the chital benefit more, since they obtain food and detect predators more efficiently. The concentrated supply of food beneath the trees causes chital to group together more densely than is usual when they are feeding. Close proximity breeds tension, and large males frequently threaten other chital competing for the fallen vegetation.

The nocturnal sambar

Sambar are large deer, standing up to 5 ft. at the shoulder and weighing up to 700 lbs. They are found in the Philippines, Indonesia, south Asia and south China. They occupy a wide range of woodland habitats but prefer open woodland on the hillsides, often near cultivated areas. In the Himalayan foothills, they live at altitudes of up to 10,000 ft. Sambar feed extensively on the foliage of low bushes.

Sambar are mainly nocturnal animals, feeding by night and hiding in dense vegetation during the day. Such elusive behavior has earned the sambar the name "invisible deer" from many zoologists, frustrated at being unable to find them easily for their studies.

Sambar are usually found in groups of five or six, comprising the mother, her offspring of the present year, the previous year's offspring and two or three other animals. In central India, males shed their antlers from March to April, and in Burma from May to June. Rutting and mating take place in November and December. In Malaysia, sambar shed their antlers throughout the year. Where there is a seasonal rut, males are more aggressive toward each other and show signs of territoriality. They do not form long-lasting harems. The spotted young are born after a 240-day gestation period.

Alarm behavior in the sambar resembles that of the chital and swamp deer. When there is a disturbance, they move toward its source with head and tail erect, giving a low alarm bark. Once the deer has assessed the situation, it usually retreats silently into the forest.

A water-loving deer

Swamp deer are medium-sized deer with long legs and elegant bodies, standing up to 4 ft. at the shoulder but weighing a maximum of only 400 lbs. They are found in north and central India and southern Nepal. Swamp deer are a grazing species, but as their name implies, they prefer a swampy environment where water is freely available. The ideal habitat consists of floodplain meadows where grasses, reeds and bamboos are available as food. Drainage and land reclamation schemes in large parts of the ranges of the swamp deer are seriously threatening their natural habitat.

The swamp deer is an endangered species, and already one subspecies, Schomberg's deer, has become extinct (in 1930). Loss of habitat is one reason for their decline; poaching is another. There is a CITES (Convention on International Trade in Endangered Species) prohibition against trade in swamp deer hides, but the total population is still small. As swamp deer breed only once a year and produce only one fawn, these animals have a slow rate of population growth and recovery. However, after several decades of decline, the populations have started to increase as a result of conservation measures. There are now more than 5000 animals in existence. About 95 percent of these are of the northern race. Numbers of the southern race are still low, but are increasing.

Swamp deer form large herds, typical of many ungulates living in open environments. For six to eight months of the year, the herds contain males, females and their young. At other times the animals separate into mother-young groups and male groups. The size of the herd increases to a maximum of 500 animals before the rut, but then dwindles to half or a quarter of this number.

The rut takes place at different times of the year, according to the area where the animals live. The timing of the rut seems largely to depend on the cycle of antler shedding and growth in male deer. For

ABOVE A male wapiti leaves the water after a bath. Wapiti live in North America, China and Mongolia, and have suffered severely from habitat destruction over much of their range.

PAGES 402-403 A group of roe deer browse in snowbound terrain. Territorial behavior among males is relaxed in winter, and deer of both sexes may gather in groups of up to 20 animals.

example, swamp deer in central India shed their antlers in May, northern populations in February or March, and those introduced into Britain shed in the late spring.

During the rut, males establish a hierarchy of dominance—the highest-ranking animal dominates those beneath it, and so on down to the lowest-ranking animal. Dominance is enforced by displays and threats, but rarely by physical conflict. The male at the top of the dominance scale does not lead the herd, or mate exclusively with all the females. However, most matings do occur between females and the higher-ranking males.

Rival male swamp deer use much less energy than the red deer to establish and maintain their position. They are therefore left in better physical condition after the rut. The males watch and follow females that are ready to mate, walking beside them and gently herding them back if they try to move away.

The endangered Eld's deer

Like the swamp deer, Eld's deer (also known by its Burmese name "thamin") lives in low, marshy country. It is of similar size to the swamp deer and has magnificent antlers that can grow to over one yard long, forming a continuous bow-shaped curve over the head. As a modification to life in marshy habitats, the Eld's deer walks not only on its broad hooves, but also on the undersides of its hardened pasterns (the part of the leg between the hoof and the fetlock).

There are three subspecies of Eld's deer, and all are endangered. One of these, the brow-antlered deer or sangai, has the dubious distinction of being the most endangered deer species in the world. In 1975, only 14 animals were still alive in a 4-square-mile area of swamp in the Keibul Lamjao National Park in Manipur, northeast India. Since 1977 sangai have been reared in zoos, and a conservation program has been started to protect the wild population. The Thailand brow-antlered deer is also at critically low levels, brought about by loss of habitat and over-hunting. The situation is a little better for the third subspecies, the Burmese brow-antlered deer, whose falling numbers are also caused by destruction of its habitat. Some are kept in zoos and reserves, and a few thousand animals are thought to survive.

The Oriental sika

The sika is found in Japan and east Asia, and could be regarded as the Oriental equivalent of the fallow deer. It is roughly the same size but weighs much less. Like the fallow deer, this species has been bred and raised in semi-captivity for many years. Both the meat and the antlers are sold, the antlers being ground down and used in medical preparations. Sika have been successfully introduced into parts of Europe, and in some areas, such as northwest England, they interbreed with indigenous red deer. A complete antler cycle in a mature male sika (from shedding to stripping of the velvet) takes 125 days and, as in other temperate species, follows a seasonal pattern. During the rutting season, sika stags utter a penetrating whistling call.

Thirteen subspecies of sika are known, many of which are in danger of becoming extinct. The Taiwan (or Formosan) sika, which used to live on coastal flats, riverbanks and low foothills, probably became extinct in the wild in 1969. Several hundred animals survive in commercial deer farms in Taiwan. One of the six Japanese subspecies, the Ryukyu Island race, may number as few as 30 animals, confined to the small island of Yakabi. Three Chinese subspecies are considered to be seriously endangered.

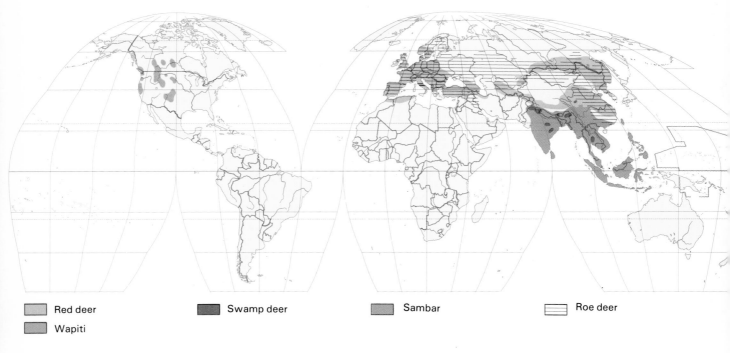

	Red deer		Swamp deer		Sambar		Roe deer
	Wapiti						

BELOW LEFT The geographical distribution of five members of the true deer family. RIGHT Pere David's deer once inhabited parts of northern China but may have died out in the wild some 3000 years ago. A captive herd was maintained in China up to the 19th century, and its descendants still survive today in European zoos and parks. BELOW RIGHT The water deer is a shy and elusive animal native to China and Korea. Its natural habitat includes reedbeds, grassland and swamps.

The adaptable red deer

Red deer are one of the most widespread and well-known members of the true deer family, and are found in Europe, Scandinavia, North Africa and central Asia. They live in woodlands and around the edges of forests throughout most of their range, but are found on open moorlands in Scotland. With 12 subspecies covering a huge geographical area, it is hard to generalize about their exact physical characteristics. However, the antlers are usually branched in a complex manner, with six to ten points on each branch. A strong neck is necessary to support the antlers, which weigh up to 20 lbs. and grow to one yard or more in length in adult males of the European species.

The size and weight of red deer vary enormously according to the subspecies. One of the largest subspecies weighs up to 660 lbs. and may have antlers 4 ft. in length, yet the Sardinian subspecies, at the smaller end of the size range, weighs less than a quarter of this amount. Several of the subspecies, including the Shou, the hangul or Kashmir stag, the Bactrian deer or Yarkand stag and the Barbarus deer, are classified as threatened in CITES (Convention on International Trade in Endangered Species) list.

Scottish deer

Much information on the social behavior of red deer has been drawn from studies made of these animals on the Scottish island of Rhum. For 10 months of the year the sexes live apart, as in fallow deer. Groups of stags, ranging in age from just-matured three- or four-year-olds to old males, are formed, and contain up to 20 animals. Within these groups there is a strong dominance hierarchy. The female groups consist of mothers and young, females without young and immature stag calves under the age of four. These herds vary in size, particularly as the young stags often move between groups.

In late September, the stag groups start to break up as the males search for females with which to breed. A stag will try to defend a herd of females from other males, moving with the females as they forage. Since the stag who is in control of such a harem will mate with its females, contests between stags for control of these groups are frequent and taxing.

Red deer stags are usually exhausted and in poor condition by the end of the rutting period. By late October and November the rut is over and males have returned to their bachelor herds. Female red deer will give birth to single young after a 230-240-day gestation period. The newborn calf is covered with hair; the coat is dappled with white spots, which fade by the time it is two months old. The calf can stand within the first minutes after birth and is able to run after just a few hours.

ABOVE The swamp deer is well adapted to life on riverbanks and in swamps. Its broad hooves spread the animal's weight so that it can walk over ground too soft for other ungulates of similar size. The deer's coat is chestnut brown, light yellowish underneath. By the early 1970s the swamp deer population stood at about 3500.

The size and condition of the antlers appear to affect the rank of an individual, although body weight and general condition are also important. A hierarchy, as found in swamp deer, exists in the bachelor herds. A series of head and antler movements, ranging from a slight flick to a savage prod, are used to establish the hierarchy in early winter.

Once established, rankings are fairly stable until March when the antlers are shed. Highly ranked stags tend to shed before younger, lower-ranked animals, so it is at these times that reversals in dominance may occur. When all the stags have shed their antlers, challenging for position in the new hierarchy is more frequent. The stags rear up on their hind legs and "box" with their front hooves, since they have no antlers with which to display.

During the rut, dominance displays and conflicts between males become ritualized, giving opponents several chances to assess each other's strength before

TRUE DEER CLASSIFICATION: 2

Cervinae

The subfamily Cervinae includes some of the most familiar Eurasian deer. Members of the group often have complex antlers, consisting of many points and branches. Four genera and 14 species are known. Recent introductions of several members of the subfamily have extended its range into Australasia.

The genus *Dama* contains the fallow deer, *D. dama*. The Persian fallow deer is normally classed as a subspecies of the fallow deer, but some zoologists consider it a separate species, *D. mesopotamica*. Fallow deer are found naturally in Mediterranean Europe, Asia Minor and Iran, and as an introduced species in the other parts of Europe and Australasia.

The genus *Axis* includes the chital, *A. axis* (also known as the spotted or axis deer) found throughout India and Sri Lanka; the hog deer, *A. porcinus*, found from northwest India through Southeast Asia to Indonesia and the Philippines; Kuhl's, or Bawean, deer, *A. kuhlii*; and the Calamian deer, *A. calamianensis*. Kuhl's deer is found in the wild only on Bawean Island, near

Java, and the Calamian deer only on the Calamian islands of the Philippines; both are sometimes considered mere subspecies of the hog deer. Chital have been introduced into many countries, including Brazil, Argentina and the USA (Florida).

The genus *Cervus* includes Thorold's deer, *C. albirostris*, of Tibet; the swamp deer, or barasingha, *C. duvauceli*, found in northern and central India and southern Nepal; the Rusa, or Timor, deer, *C. timorensis*, of the Indonesian islands; the sambar, *C. unicolor*, of the Philippines, Indonesia, south Asia and south China; Eld's deer, *C. eldi*, of Southeast Asia, Manipur (eastern India) and Hainan Island; the sika, *C. nippon*, of Japan and east Asia; the red deer, *C. elaphus*, of Europe, North Africa, Asia Minor and central Asia; and the wapiti, *C. canadensis* (known in North America as the elk) of the northwestern USA, China and Mongolia. Several of these species have been introduced in other parts of the world.

The genus *Elaphurus* contains a single species, Pere David's deer, *E. davidiensis*. Once found in the lowlands of northeast China, it now only survives in captivity.

starting to fight. A challenger will first approach a harem-holding male. Both animals then roar at each other for several minutes. Sometimes, the greater roaring ability of an older male will deter a contender from carrying on with his challenge.

Trial of strength

The next stage is called the "parallel walk," as the deer pace along the edge of a boundary, eyeball-to-eyeball, until one suddenly turns to face the other, lowering his antlers. In hilly terrain one animal usually tries to gain the advantage of higher ground, since the contest now involves the locking of antlers as contestants try to push one another backward. In most cases, the battle is a test of strength, a stronger animal usually succeeding in pushing a weaker one backward. However, it is also true that experienced stags may have some advantage over younger challengers in their fighting style and tactics. The loser retreats at a brisk trot, often pursued for some distance by the winner.

While the harem holder has been fighting, his females often wander away or are lost to a third stag. Males are reluctant to fight, probably because fighting uses so much energy and is a risky business. (Up to 23 percent of males are injured in the rut, and up to 6 percent permanently injured.) Instead, many contests are settled by the preliminary roaring match.

Unlike stags, hinds do not have a rigorously enforced dominance hierarchy. However, on Rhum, some forms of female dominance have been found, with strong hinds passing on their higher status to their female offspring. While most females will breed several times during their lives, only mature males who can defend harems are likely to breed. For this reason, it seems that female red deer allow male offspring to suckle for longer before weaning than they would allow female young, in order to build up the calf's strength. A mother who has just weaned a male calf will have to wait a longer time before she is able to mate again, suggesting that she has devoted much of her energy to her calf's early development.

Fawns are usually left lying concealed in the grass while the mother wanders away to feed, and are not suckled more than once every five to six hours. However, deer milk is very rich and nutritious, containing three to six times as much fat as cow's milk and about twice as much protein.

ABOVE Mule deer can survive for several days without drinking water, since they obtain moisture from succulent plants. During the winter they gather in sheltered valleys, forming large herds.

Gathering a harem

During the breeding season each red deer stag tries to gather a number of mature hinds into a harem. Some of the hinds may have young from the previous season, but the stag will drive away any immature deer that are already a year old. Other stags are certainly not tolerated in the neighborhood in case they lure away some of the hinds for their own harems. To prevent this from happening, the stag is forced to put on continual displays to intimidate possible rivals and indulge in tiring chases or even battles. These usually start with a roaring contest (the deep, loud roaring is known as "belling"). The stag which roars most frequently in a given time—the rate of roaring—often wins the day.

Even if there are no other males in the vicinity, the stag will not tolerate disobedience on the part of his hinds. He is constantly rounding them up and keeping them in line. If a hind tries to leave the harem, he will overtake her and block her path, then attempt to assert his supremacy by stiffening all his muscles and adopting a curious swaying walk. He bares his

THE WHITE-TAILED DEER
— A DEER OF THE AMERICAS —

The white-tailed, or Virginian, deer and its subspecies are found in both North and South America, from southern Canada south to Peru and northern Brazil. Most studies have been conducted on the North American subspecies in their woodland habitats or in semicaptive herds.

One study, from the Crab Orchard Wildlife Reserve in southern Illinois, USA, produced the following picture of the social organization of the animal. As in many deer, the sexes live separately. Bucks are usually solitary from autumn until early spring, but form associations between spring and autumn. These bachelor groups are small—80 percent in the study contained fewer than four individuals—

and are usually composed of unrelated males. Females give birth to between one and three young in early summer about seven months after mating. They give birth alone, and from May or June until early autumn lead solitary lives while they tend the fawns.

The young, like the adults, are well camouflaged. Their brown fur and white spots blend in cleverly with the sunlight-dappled leaves. In the beginning, the fawns are left in thick cover, and the mother returns only to suckle them. When they are a few days old they are able to follow her. After six weeks the fawns are weaned, and although agile enough to evade being caught by humans, they stay close to the mother until they are two years old.

Between October and December, female yearlings rejoin their mothers and the young of the current year. Sometimes, under very good conditions, these yearling females will already have bred and produced their own offspring. This may bring the number in the group up to about eight. The family groups stay together throughout the winter, but break up in May and June.

Signals by scent

Scent cues are important forms of communication in largely solitary animals, and white-tailed deer show a range of marking behaviors. Glands on the metatarsal bones, below the ankle joints on the hind legs, as well as interdigital glands beneath the hooves, are used to leave scent trails.

Urine is often mixed with the glandular deposits on ground that have been scraped clear with the hooves. Males may scent-mark and defend areas surrounding a female during the breeding season. Bark-stripping and marking with secretions from the facial glands also occur in this species.

White-tailed deer browse on tree leaves and low plants for most of the year. They normally need from two to ten pounds of food each day—this includes acorns, buds, beechnuts, shoots and other vegetation. Studies have found that seasonal changes in the diet do occur. In some oak forests during autumn, acorns make up 80 percent of the diet. In winter, when the quality and quantity of suitable foliage are reduced, the white-tailed deer's appetite lessens, an adaptation that saves the animal from spending valuable energy and time looking for food. As soon as spring arrives and new plants are available, the deer's appetite returns to normal.

A

B

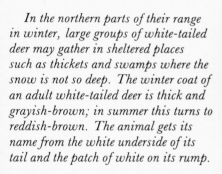

In the northern parts of their range in winter, large groups of white-tailed deer may gather in sheltered places such as thickets and swamps where the snow is not so deep. The winter coat of an adult white-tailed deer is thick and grayish-brown; in summer this turns to reddish-brown. The animal gets its name from the white underside of its tail and the patch of white on its rump.

LEFT A stag white-tailed deer in its typical waterside habitat. The range of the species runs from North America south to Brazil.
ABOVE A stag lowers its head submissively before a rival (A); rival stags face each other sideways (B).
TOP RIGHT Between one and three spotted fawns are born to female white-tailed deer.
CENTER RIGHT The white tail is used to give alarm signals. Scent-marking is another important means of communication for white-tailed deer.
BOTTOM RIGHT A stag clearly shows the layer of "velvet" which covers deer antlers during their annual growing period.

Chital

White-tailed deer

Swamp deer

Sika

Eld's deer

RIGHT Pampas deer have suffered from hunting and from competition with domestic cattle for grazing land. One subspecies is now the most endangered deer in all South America— only a few hundred individuals still survive in the wild.

canine teeth in a threatening manner, at which point the hind usually decides that her best course of action is to return to the harem. As she makes her way back, the stag dashes up behind her, prodding her with his antlers and uttering the short coughing cry used by red deer to signify recognition, challenge or threat.

The stag checks his hinds continually to see whether they are ready to mate. He does this by smell, monitoring their resting places, droppings, urine and prints—and if he detects the scents of a hind going into season he reacts to them with the characteristic "flehmen" response—throwing his head back and curling his lips away from his front teeth. Having decided that a hind is ready to mate, the stag does not approach her in slow, seductive fashion, but chases after her briskly—although not aggressively. He holds his neck horizontal, with his muzzle pointing upward in the flehmen position. If the hind is not receptive, she will move away and try to lose herself within the herd or run around the back of it. If she is receptive, on the other hand, she waits with her back slightly hunched, her tail raised and her rump flash displayed. As soon as she does this the stag will mate with her.

Royal deer

These days the idea of animals surviving in zoos and parks when they have died out in the wild is depressingly familiar, and such a situation is usually regarded as the prelude to extinction. But one species in the true deer family may have survived in this state for centuries. Pere David's deer, a large animal with a long tail and tall, branched antlers, was discovered in the walled Imperial Hunting Park near Beijing (Peking) in 1865 by the Jesuit priest Pere Armand David. Subsequently, some of the deer were donated to zoos in Germany, France and Great Britain— fortunately, as it turned out, for the Chinese royal park was flooded in 1895, drowning most of the deer. Those that survived were killed during the Boxer Rebellion of 1900. After this the zoo animals in Europe represented the entire stock of Pere David's deer, but luckily the species proved prolific in captivity and was bred with great success by the Duke of Bedford on his Woburn Park estate in the English Midlands. Descendants of the Woburn herd now flourish in zoos and wildlife parks throughout the world, and the species has also been reintroduced to its native China. By the early 1980s the deer's world-wide population had reached over 1000 animals.

Water deer

The water deer (or Chinese water deer) of eastern Asia is a distinctive species that has been classified in its own subfamily, the Hydropotinae. Like the musk deer it has no antlers, but it does have protruding tusk-like upper canine teeth that, in males, may grow up to 3 in. long. These deer weigh about 20-35 lbs. and are built like the muntjacs and other deer typical of scrub and dense woodland. They have highly developed hindquarters and short, humped backs. They are territorial, basically solitary animals. Two subspecies are known; one is distributed throughout the wetlands of the Yangtze River in China, while the other lives among the swamps and grasslands of Korea. Both have been introduced successfully in some areas of France and Britain, where they live in parks and in low-lying wetlands such as the East Anglian fens. Water deer are extremely shy creatures that keep to vegetation cover and are difficult to spot.

ABOVE **The little red brocket is a deer from forested parts of South America. It is** an elusive animal, active by night and hiding by day amid thick vegetation.

The European roe deer

Although described in great detail in various treatises on hunting over the past hundred years or so, the roe deer was not studied in depth by naturalists until the late 1960s. Until then our knowledge of its behavior was relatively sketchy. It is a medium-sized deer, rarely exceeding 55 lbs. in weight. The European subspecies is found throughout Europe from Greece to Scandinavia, and also occurs in Asia Minor. Its natural range extends from the Ukraine and Poland in the east, to Spain and Portugal in the west. As a result of this large area of distribution, it occurs in a number of forms, formerly all regarded as separate subspecies. In England, Siberian roe deer were introduced about 100 years ago to replace the original native roe deer that had died out over most of the country.

The roe deer lives in woods and forests. Like many woodland animals it has little stamina, and to avoid being chased by enemies, it disappears quickly into the forest undergrowth when threatened. It has short antlers, a conspicuous black nose and a short, well-proportioned body that varies from reddish in summer to gray-brown in winter. Its tail is almost non-existent, but it does have a large rump patch, not so noticeable in summer when it is pale buff, but very prominent in winter when it is startlingly white. The white rump hairs are fluffed out like a powder puff when the deer is alarmed.

In well-wooded areas, roe deer tend to be solitary animals during late spring and early summer. There is no herd structure, as there is among red deer, and the males do not gather harems. Monogamy is the rule, with males and females forming pairs during the midsummer mating season. They stay together through the autumn and into winter, often accompanied by one or two immature females born the previous year. One to three fawns are born in the spring, deep in the undergrowth, and when they are about two weeks old the doe brings them out to join her on foraging expeditions.

On open ground, in early winter, these small groups may occasionally gather together to form large assemblies of up to 20 animals. This is possible because the buck becomes less jealous in the period immediately after his antlers are cast in November. He enters an "indifferent" stage when he ignores other roebucks. The level of male hormone in his bloodstream is low, and his instinct for sexual rivalry is dormant. It is during this phase that the new antlers start to grow.

In contrast to fallow, red, sika and muntjac deer, which shed their antlers in spring or early summer and regrow them during summer, roe deer lose theirs in late autumn and regrow them ready for a midsummer rut. During late winter, the level of testosterone—the male hormone in the blood—starts to rise. This has two effects: it makes the antlers harden and lose their velvet, and it makes the buck much more aggressive toward potential rivals. The buck's antlers are fully grown by spring.

Bucks in battle

If two bucks meet in the forest they will try to intimidate one another. A roebuck does this by approaching his rival head-on, then stopping and turning his head slightly to one side while keeping his chin raised. This haughty attitude often has the desired effect, for the other animal will visibly shrink in stature as it adopts a submissive stance. Sometimes the dispute is not so easily settled, and both animals resort to threats. They then lower their heads with antlers

ABOVE Timid and solitary, the southern pudu lives in the deep forests of the lower Andes. It is the smallest of all the true deer, little larger than a Jack Russell terrier. Yet in anatomy it closely resembles the majestic red deer of Europe and Asia. As with so many deer, it has suffered from widespread and indiscriminate hunting.

upturned and engage in a brief tussle, which is often over so quickly that it can hardly be detected. A roebuck's antlers are short but sharp, and can be deadly—a number of people have suffered fatal injuries from them—but most fights between bucks end in the flight of one of the combatants.

These skirmishes serve to establish the boundaries of each buck's territory, which usually covers some 12 to 50 acres. The defended areas are patrolled quite regularly, and the boundaries are marked to show other roe deer that the area belongs to one individual.

Marking territories

Boundary marks are of two types: scent-marks and visual signals. Scent-marking is carried out using secretions from the adult buck's facial glands. These glands are especially abundant on the forehead and at the base of the antlers. The buck rubs his head against bushes, smearing them with the secretion which carries his personal scent. Scrapes made on the ground with the feet act as visual boundary markers, but they also work as scent-marks because they contain secretions from glands between the buck's toes. Roebucks will also thrash at saplings with their antlers, stripping the bark as a visual sign and anointing the wood with scent from the glands on their forehead.

As the stags go through their territorial phase in spring, the females go off into the undergrowth to give birth. It is common for only one of the young to survive the first winter. The bond between mother and offspring is weak. For three or four weeks the fawns are left on their own, lying among the thick vegetation, and their mothers visit them only to feed or clean them. This system offers an effective means of defense in thick woodland, for the fawns are well camouflaged with spotted coats, and they do not draw attention to themselves. They are not as vulnerable as they look, and it is a mistake to "save" fawns found alone in the forest since they are never actually abandoned. The fawns' spotted coats are lost when they are a year old.

In northern Europe the rut takes place at the end of July and beginning of August. When a mature doe enters a buck's territory he will court her. This can be an involved affair, with the male selecting an open space around a tree or a bush, scraping the ground and marking the surrounding area, and then chasing the doe around and around in a ring or figure-eight until she lets him mount her. The couple will often wear a prominent circular track in the forest floor, known as a "roe-ring." The pair stays together for the rest of the season. When the buck leaves, he may move several miles away from his area.

The roe deer is one of the few species that have gradually increased in numbers over the last few years. This is partly because many of its natural predators (such as wolves and lynxes) are in decline, but also because it has adapted to the human landscape and is able to colonize areas that are under cultivation—as long as it can retreat to nearby woods. The roe deer is also important as game in many parts of Europe.

The Asian roe deer—the Siberian and Chinese subspecies—are larger than their European counterparts and are quite different in appearance. The Siberian variety may weigh up to 110 lbs. and has sturdier, more branching antlers. They also behave differently; Siberian roe deer are highly gregarious and may form herds of over 100 animals during the winter.

413

ABOVE **A bull elk ruminates on the riverbank, his antlers still bearing tattered remains of their recent cloak of velvet.**

BELOW **The map shows the world distribution of several of the true deer, along with the musk deer and the Asian chevrotains.**

White tails, black tails

The Virginian deer, or white-tailed deer, is widely distributed from southern Canada to Brazil and has also been introduced in Scandinavia, New Zealand and Cuba. There are 38 subspecies, which vary considerably in form: the more northerly varieties tend to be larger (up to 430 lbs.), while those subspecies further south are medium-sized and small, such as the Florida Keys white-tailed deer, which usually weighs less than 55 lbs.

The mule deer is another American species and is similar to the white-tailed deer. The antlers are forward-branching in both species. Mule deer are found only in western North America and Central America. There are a number of subspecies (seven are generally recognized), ranging in weight from 100 lbs. to over 450 lbs. They are found mainly on prairie grasslands, shrublands and dry open forests, and range further into the mountains than the white-tailed deer. In summer, mule deer can be seen on upland slopes, browsing on the fresh growth of trees such as pine and aspen. After the autumn rut, they migrate down to the shelter of valleys. Unlike white-tailed deer, mule deer do not adapt well to environments modified by man, and many subspecies native to highly populated areas are endangered.

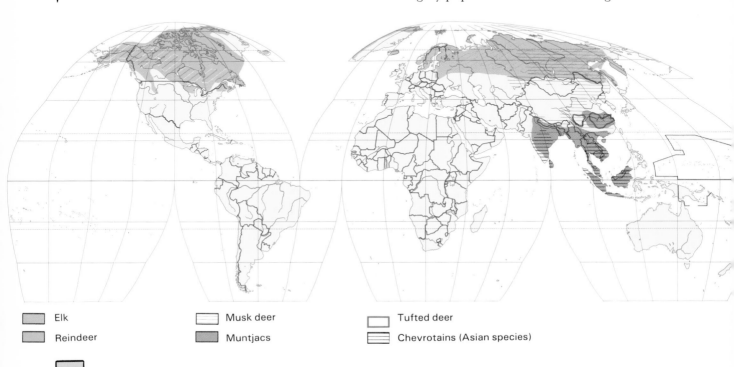

Elk

Reindeer

Musk deer

Muntjacs

Tufted deer

Chevrotains (Asian species)

ABOVE A temporary liaison: two bull elk keep company for a while as they graze in the open country at the edge of a conifer forest. Elk tend to be solitary creatures and never form groups for long; during the winter their survival depends on a secure food supply, and that means driving out any other elk that would compete for their scanty ration of leaves and small plants.

RIGHT A dominant male (A) asserts his authority over a subordinate (B). The latter adopts a submissive posture with its head held low.

Black-tailed deer

One of the best-known subspecies of mule deer is the black-tailed deer of northern California and Oregon. It is a species with a complex social organization. At least two forms of territorial behavior are known: maternal and group. A doe will aggressively defend an area of 100-200 ft. in diameter centered on the spot where her fawn is lying, as long as it stays still. Curiously enough, if the fawn is active, the territory extends over an area of only 16 feet. Some naturalists have concluded that if the fawn moves, the territory moves too; others believe that the territory is a distinct geographical unit that does not change.

The black-tailed deer live in clans like red deer, and the stags defend the group territory against other mule deer. The males have a wide range of aggressive techniques: not only do they threaten one another with their antlers and charge at one another, but they will often kick with their forefeet. When one animal concedes defeat, the dominant animal will emphasize his superiority by placing a foot on the loser, which cowers on the ground.

South American deer

The marsh deer (or guasu pucu) and pampas deer of South America are near relatives of the mule deer. The marsh deer, the largest native South American deer, is about the size of a medium-sized European red deer, weighing 220-330 lbs., and has evolved into a highly specialized animal. Its physical structure has adapted to life in the swamps of southern Brazil, Paraguay, southeast Peru, Bolivia and northeast

Elk

Sambar

Wapiti

Reindeer

416

The natural range of the true deer covers most of Europe, Asia, North America and South America.

Argentina. It has long, slim legs with elongated, separable hoofs that are joined by a membrane that allows the deer to walk on soft, marshy ground. Its coat is rough and thick, with dark hair on the legs, and its antlers are subdivided in such a way that each tine (branch) bears a fork. Like all highly specialized animals, it has become equipped for one particular habitat only, and because of this it is vulnerable to the environmental changes that are taking place throughout its range.

The smaller pampas deer (weighing only 65-90 lbs.) is also under threat, and one of its three subspecies—the Argentine race—is particularly at risk. Numbers dropped as low as 100 by 1975, but with protection rose to 400 by 1980. Their decline came about not only due to hunting, but also because much of the pampas grassland in Brazil and Argentina has been taken over for farming and cattle rearing. This deer tends to be monogamous in its habits, like the European roe deer, and as a result it often lives in small family groups.

Driven out by agriculture

The Peruvian huemul is a medium-sized deer with small, forked antlers and large ears. It is listed as a vulnerable species, despite the fact that it lives in relatively inaccessible parts of the high Andes from Peru to Argentina, at altitudes of 10,000 to 16,500 ft. These animals are suffering the same fate as the vicuna: they are being driven away or wiped out in the less harsh parts of their range and their places taken by the domestic animals and crops of mountain farmers. However, in some localities their numbers may be increasing; since they share their habitat with vicunas, they benefit from protection given to these animals in reserve areas. Protected sites in which they occur include Lauca National Park in Chile and the Ulla Ulla Fauna Reserve in Bolivia.

The Chilean huemul is very similar to the Peruvian huemul, and it is even more at risk. Today it is largely restricted to the national parks of Monte Lanin, Nahuel Huapi, Los Glaciares and Los Alerces in Argentina, and the Rio Simpson National Park in Chile. The total population in the early 1980s was less than 2000.

The brockets and pudus are small deer from South and Central America, with tiny antlers. The brockets number four species, each weighing about 45 lbs., apart from the dwarf brocket that weighs a little over 20 lbs. In spite of their small size, and the fact that their antlers are reduced to two simple spikes, they are more closely related to the large red deer than to the small deer of the Old World (such as the water deer and the muntjacs). The same applies to the two species of pudu, which also have very small antlers. The southern pudu is the smallest member of the deer family, weighing 18 pounds at the most.

Giant of the north

The elk is by far the largest of all the deer—males stand over six feet high at the shoulder and weigh up to 1750 lbs. (a bulk roughly one hundred times that of the southern pudu). Its spectacular antlers are branched and palmate—that is, they are flattened and outspread like the palm of one's hand, with the finger-like tines curving upward. The elk has a long nose, a drooping lip, and a hump at the shoulders. Its long limbs have broad, splayed hooves that spread the animal's weight as it walks over the muddy or swampy ground typical of its natural habitat—the great coniferous forests of the north. It feeds on foliage stripped from trees, such as willows and sallows, and on plants growing on the forest floor. Aquatic plants such as horsetails and pondweed may also be eaten, and the animal will wade in water up to its belly for hours if the temperature is not too low. It can swim well and has even crossed sea straits in search of new grazing grounds.

The elk is found in parts of northern Europe, Siberia, Manchuria, Alaska, Canada and parts of the United States. In North America it is known as the moose. In Roman times elks seem to have been widespread in Germany, and Julius Caesar himself described animals living in the ancient Hercynian Forest running from the Rhine to the Carpathians. However, by the 17th century they had died out west of the Elbe River, and during the first half of the 20th century it was feared that they might disappear from Europe altogether. Fortunately, since the end of World War II, the European elks have gained a certain amount of ground, and they can now be found in parts of Germany, Poland, Norway, Sweden and Finland.

Elk tend to be solitary animals, although they do not live in the thick, bushy habitats occupied by most solitary deer. Their unsociable habits can probably be explained by the relative scarcity of food in the northern forests in winter. If the elk population is spread out, there is less competition for food in each area, and the animals stand a better chance of surviving the harsh conditions. It has been estimated that at least 50 percent of adult elk live alone; 25 percent forage in pairs, and the remainder live in groups of three or more. These figures do not take into account any young that might accompany adults. Groups tend to be temporary, because individual elk become attached to their own home ranges and will abandon their companions rather than stray into unfamiliar territory. The young do not stay long with their mothers and often leave after a year, young males usually leaving before females.

Aggressive mothers

Mature female elk are extremely aggressive toward their own species. During the winter, when they are accompanied by their young, they will not tolerate the approach of any other elk, regardless of age or sex. This aggression could well be a tactic to defend the winter food supply in the area, which may be only sufficient for one animal and its young. The greatest threat to an elk's survival is not predation but food shortage in winter, and this means that the main enemy of an elk is another elk.

Immature elk tend to disperse from the areas where they were born and raised, but once they reach adulthood they stay within stable home ranges containing summer and winter quarters that are revisited year after year. Each elk occupies an area of 750-1500 acres and tends to commute between open forest and rivers or lakes every day. A well-developed code of ritualized combat maintains the boundaries between neighboring territories, and, assuming there is sufficient room for all the elk to stake out a good-sized territory, the system works well.

Occasionally an elk population may grow to the stage that there are too many animals for the resources available. This is what happened at Isle Royale, a wilderness island in the north of Lake Superior on the border between Canada and the USA. A group of elk reached the island in 1904, and in the course of 25

TRUE DEER
CLASSIFICATION: 3

Odocoilinae

The deer subfamily Odocoilinae comprises nine genera and a total of 15 species. They are mostly New World animals, but some species occur over Europe and Asia.

The white-tailed deer, *Odocoileus virginianus*, is a widespread America species found in conifer woods, swamps and river valleys from Canada to Brazil. The mule deer, *Odocoileus hemionus*, lives in Central America and western North America. One of its 11 subspecies is the black-tailed deer of California and Oregon. The roe deer, *Capreolus capreolus*, has a wide range from Europe across northern Asia to China. The largest of all deer, the elk, *Alces alces*, ranges from northern Europe through Siberia to China, and occurs over much of North America, where it is known as the moose. The reindeer, *Rangifer tarandus*, lives in the forests and sub-Arctic tundra of the far north in Scandinavia, Russia, Siberia and North America. In North America it is known as the caribou.

The remaining ten species are native to Central and South America. The marsh deer, *Blastocerus dichotomus*, is found in marshes and savannahs from central Brazil to northern Argentina, and its range is similar to that of the pampas deer, *Ozotocerus bezoarticus*. There are two species of huemul: the Chilean huemul, *Hippocamelus bisulcus*, of the high Andes in Chile and Argentina, and the Peruvian huemul, *H. antisensis*, of the high Andes from Ecuador to northern Argentina.

The two species of pudu occupy the lower Andes: the southern pudu, *Pudu pudu*, lives in Chile and Argentina, and the northern pudu, *P. mephistophiles*, is found in Peru, Ecuador and Colombia. Four deer belong to the genus *Mazama*, the brockets: the red brocket, *M. americana*, and the brown brocket, *M. gouazoubira*, range from Mexico south to Argentina, living in mountain thickets; the little red brocket, *M. rufina*, is found in Venezuela, Ecuador and southeast Brazil; and the dwarf brocket, *M. chunyi*, occurs in the Andes of northern Bolivia and Peru.

years they reproduced at such a rate that they achieved an alarmingly high population density of over five animals per 250 acres. Inevitably this situation led to starvation each winter, because the vegetation cover was almost completely destroyed, and many elk died every year.

Then, in 1950, a group of wolves reached the island over the ice, started preying on the weakened elk and reduced the population to a more appropriate size. The vegetation regenerated and, with an adequate food supply, the elk regained their strength. Since an elk in good health is a match for any wolf, the two species now live in perfect equilibrium, to the mutual advantage of both. The wolves live by preying on old and weak elk (among other animals), and this predation maintains the elk population at a level that ensures sufficient food.

Reindeer of the tundra

The reindeer is an animal of the far north. It is found all around the Arctic Circle from Norway, through Sweden, Finland, European USSR, Siberia, and Alaska, to Canada and Greenland. In winter, reindeer keep to the edge of the conifer forests, but in summer they range over the bleak, thinly vegetated Arctic tundra. Despite their remote habitat, many

TOP Scraping through the snow to the grass beneath, a herd of wapiti scratch a living in the foothills of the Rocky Mountains. Closely related to red deer, wapiti were once widespread on the plains of North America, China and Mongolia, but they have been driven into wilder country by the spread of agricultural land. ABOVE Its wet flanks gleaming in the low sunlight, an elk picks its way across the valley grassland. Elk are the largest of the deer; a male may weigh up to 1750 lbs.

419

ABOVE **A bull reindeer (foreground) forages with two females. Unlike all other deer, both sexes have antlers—although those of the male are heavier and more complex. The female probably uses hers to** defend her grazing rights on the thin northern pasture, and to defend the young against predators such as wolves.
BELOW **A female reindeer lowers her head as she gently guides her calf from behind.**

reindeer are semidomesticated. In the past, the nomadic Lapps of northern Scandinavia followed the herds on their migrations to and from the tundra, and used them for food, transport and to provide clothing. Today the Lapps have abandoned many aspects of their traditional way of life, but they still farm the reindeer, and they still follow them when they leave the security of the forests for open tundra.

The vast geographical range of the reindeer has led to the separate development of several subspecies (up to nine or more, according to some scientists). These subspecies vary greatly in size from race to race. The smallest weigh from 130 to 175 lbs., but individual specimens of the biggest subspecies may grow to 600 lbs. or more.

The American reindeer, known in North America as the caribou, is the largest subspecies. Unique among deer, both sexes bear antlers. It has been suggested that the females need their antlers to defend their grazing rights from competing males and females when feeding in large herds on the thinly vegetated grazing lands. The animals use their antlers and hooves to scrape away the snow to reach the vegetation beneath it. Calves feed from the open areas made in the snow by their mothers.

Reindeer are well adapted to the rigors of life in the harsh northern climate. They are long-legged animals with large, spreading raft-like feet that help to prevent them sinking into the snow. Their hooves make a characteristic clicking noise as they walk. Their bodies are well covered with thick hair, as are their nostrils and muzzles—a valuable adaptation for a grazing animal that often has to dig through the snow to find food.

THE SHY WOODLANDERS

Social life

Wild reindeer gather together in great herds. In North America, caribou used to migrate in herds thousands strong. They still move around in large herds, although their numbers have dwindled somewhat over the years. However, they are not always found in herds. During the summer, males are usually solitary. When the young are born, males and females separate, and the males live either in small scattered groups or on their own.

Female reindeer give birth to their young in June. The mother remains with her offspring for two to three days, but by the fourth day she begins to seek the company of other mothers with their offspring, forming small "nursery herds." Later in July, biting flies begin to be a real problem, and the herds tend to break up as each animal tries to find some shade, where the flies are not so troublesome. If this is not possible, they often stay on the move to avoid becoming a sitting target for the insects.

The onslaught of the flies stops in September, as the mating season approaches. At this stage the reindeer gather in small groups of a few dozen individuals, often dominated by one adult male who asserts his authority over both females and other males. These are not harems, for the males rarely stay with one particular herd, but move from group to group, mating with females when and where they can.

Rut behavior

The rut is marked by distinct patterns of behavior. Females often become very aggressive. Indeed, they always come off best in encounters with sub-adult males. For their part, mature males advertise their status with a harsh, guttural sound, rather like the roar of a red deer stag. Another form of behavior, known as "wither rubbing," has similarities with that of the American white-tailed deer. The male places his hind feet close together and moves them under his body. As he does this, he urinates, then rubs the scent glands on his feet against one another. In another posture, which often occurs at the beginning of an attempt to

EURASIAN DEER

Deer populations have suffered a long history of decline. For many centuries their greatest enemies have not been predators or natural diseases, but the twin destructive forces which threaten so many wild animals today—habitat loss and hunting.

In Europe and Asia the history of decline commenced thousands of years ago. Most deer are woodland animals, and all over Eurasia woodlands were gradually cut down to make way for the spread of settlements, pastures and cultivated land. While their habitats dwindled, their need for cover increased—all but a few species were the quarry of hunters who killed more and more deer for meat and for sport. Only in recent years has the need for conservation received strong and widespread support. Yet the need has dramatically deepened; the pressures on deer are now more pervasive than ever.

The red deer once occurred throughout Europe, the Middle East, northern Asia and northernmost Africa. Now its range is greatly

fragmented, and several of its 12 subspecies are endangered. The hangul, or Kashmir stag, declined to as few as 500 animals, and the Barbarus deer—the only deer native to Africa—survives only in a small area in the Atlas Mountains. The range of the fallow deer has also steadily contracted. It became extinct in Greece during the 19th century, and the Sardinian population died out in the 1950s. The Persian fallow deer survives only in small numbers on the Iran-Iraq border.

Numbers of the swamp deer of India and Nepal have declined both through hunting and loss of habitat, especially through losses arising from land drainage. All three subspecies of Eld's deer are endangered, two of them being extremely rare: the brow-antlered deer, or sangai, and the Thailand brow-antlered deer. The brow-antlered deer numbered only 14 in the mid-1970s. The sika is also endangered in many of its native areas in East Asia, and its subspecies living in the Japanese Ryukyu Islands may have been reduced to no more than 30 individuals.

421

mate with a female, the male freezes his body, keeping his head and neck slightly lowered and eyes half closed. He may stay like this for some time before making any further move.

Males rarely fight openly with their rivals and, although they may threaten one another and make much of their social status, several males can remain within the same herd without coming to blows. This is particularly true when there are more females than males, and the largest herds usually prove to contain a high proportion of females.

The herding behavior of the reindeer, which echoes that of other gregarious ungulates such as the wildebeest, has great value as a defensive strategy against predators. There is safety in numbers, and wolves tend to avoid attacking the main body of the herd and concentrate on cutting out old or weak stragglers. This saves many reindeer from falling prey to the wolves. It also ensures the health of the herd as a whole: diseased and malformed individuals are quickly picked off by the predators and have little chance to infect others or pass on their weaknesses by breeding.

Herds on the move

At the first sign of bad weather, especially the first heavy snowfall, reindeer living near forested country move into cover. Reindeer on the open Arctic tundra migrate, but the direction they take depends on where they are. The animals do not always move south, as one would expect, because some subspecies of reindeer prefer to winter on open ground: in southern Labrador open ground is to be found further north.

The animals group together for the long migration from the summer pastures to their winter quarters. The vast herds that form appear to be led by the more experienced reindeer who know the routes and traditional paths. Once the animals have arrived at the wintering grounds, they may drift apart or stay together, depending on the food supply and weather

RIGHT On the fringes of a northern forest, a group of reindeer scrape through the autumn snow to find food. The magnificent antlers of the male are fully grown, and the velvet is beginning to split and fall away. Revealed beneath the velvet are the remains of the blood vessels that nourished the antlers when they were growing. The deer, irritated by the peeling velvet, will thrash the antlers against trees and bushes to strip off the rest of the covering in preparation for the rut.

conditions. But in spring they form up in a herd again to return to their summer quarters. The females and young move off first, to be followed later by the adult males. Naturalists still do not know whether these animals are guided by a well-developed sense of direction, an ability to recognize landmarks, or some kind of "race memory."

When they are out on the tundra, the reindeer feed largely on the lichen *Cladonia rangiferina*—popularly known as reindeer moss—which grows abundantly in sub-Arctic regions. Unfortunately, lichens can absorb radioactive particles that are present in the air, and since each reindeer eats a great deal of lichen, the pollutant is concentrated in their bodies.

Radioactive reindeer

After the accident at the Chernobyl nuclear power station in the Ukraine, USSR, in April 1986, which released a radioactive cloud into the atmosphere, the lichens in northern Scandinavia became heavily contaminated with radioactivity. Reindeer meat became impossible to sell because it was so highly polluted. The long-term effects on the reindeer have yet to be fully assessed, but it is hoped that once the animals have eaten their way through the polluted vegetation, the problem will begin to subside.

Presently, about three million reindeer exist in domestication around the world. This compares with a wild population of no more than two million.

UNDER THREA

NEW WORLD DEER

Though the history of land clearance and overhunting is much shorter in North and South America than it is in Eurasia, many species of deer in the New World are now under threat; in modern times, the pace of their decline has been rapid. Before European colonization, the human population of the Americas had little effect on its deer population. Habitats were left largely undisturbed, and hunting was carried out for subsistence alone.

With the huge increase in human numbers in the New World over the last three centuries, all this changed. Not only have vast areas been stripped of their natural vegetation, but hunting took on a new character—no longer were deer killed merely to stave off hunger; they became an object of sport, and their meat, hide and antlers became a source of profit.

Before European colonization, the North American reindeer (caribou) population stood at some 3.5 million—now less than a third of that number survive in the wild. Some local populations face strong pressures from overhunting and from the spread of industrial oil installations. Huge numbers of white-tailed deer are still hunted under license in Canada and the USA. Though their overall numbers remain large, some subspecies are endangered. The key deer lives only on some of the islands that make up the Florida Keys. Its population in the late 1970s stood at some 300-400.

The pampas deer of South America has suffered over most of its range through habitat destruction and hunting. By 1980 the Argentinian population stood at about 400 animals. Though the pampas deer is legally protected throughout its range, hunters know they are unlikely to be apprehended—laws are only effective if they can be enforced. Widespread poaching has also depleted numbers of the marsh deer, as have pressures from land drainage and land clearance for agriculture.

Though the Chilean huemul occupies remote parts of the Andes, it too has faced decline because of human activity. Hunting, habitat loss and competition from domestic animals have combined to reduce its population to no more than 2000, distributed over the southern Andes of Chile and Argentina. Its relative, the Peruvian huemul, is also declining steadily in most parts of its range.

The northern pudu, which lives in the equatorial region of the Andes, is widely hunted and is particularly threatened by the destruction of its habitat during forest fires. Its relative, the southern pudu, is also pursued by hunters, but so far its population is not considered to be under immediate threat. If the twin pressures of habitat loss and hunting continue unabated, that situation is sure to change. Then the southern pudu will join the long list of deer from around the world that have the misfortune to be classed as endangered species.

A HEAD FOR HEIGHTS

With its long neck, the giraffe is uniquely adapted to browsing the treetops. Its relative, the okapi, uses its elongated tongue to grasp leaves lower down

The giraffe is one of the most highly specialized of the ungulates, superbly adapted for browsing the treetops. It is the tallest of all the mammals, and its long legs and long neck enable it to reach the foliage growing on the upper branches of trees, over five yards above the ground. No other large, ground-dwelling animal can do this (although elephants will push trees over to get at the foliage if more accessible food is not available).

The ancestors of the giraffes were short-necked animals that probably lived in forests, eating the foliage off the bottom branches of trees and low-growing shrubs. With the expansion of the African grasslands in prehistoric times, trees became scarce over large areas. Tall animals that could reach more of the tree foliage gained a distinct advantage over shorter animals, and over the years, the length of the giraffe's neck and legs increased.

Short-legged survivor

While the giraffes responded to the changing environment over much of Africa, most of their primitive short-legged, short-necked relatives died out. Most, that is, but not all—for the giraffe family has one other surviving member, the okapi.

Judging from the fossil remains of its ancestors, the okapi has changed very little over the last 20 million years or so. About 5-6 ft. high at the shoulder, it weighs up to 550 lbs.—less than a quarter of the weight of an adult giraffe. Its head is like that of a giraffe, but is mounted on a short neck, and its general proportions are similar to those of a mule. It is a colorful animal, with horizontal zebra-striping on its legs and rump, and dark hair, with a reddish or purplish sheen on its back. Its black tongue is so long that the animal can use it to clean its eyes, though its main task is to pull foliage off branches. The male also has a pair of short giraffe-like horns covered with skin and hair.

Okapis inhabit the rain forests of northern Zaire, and where the habitat is suitable they may live at population densities of two to four animals per square mile. They prefer areas where watercourses, clearings and glades allow light to penetrate through the high forest canopy, encouraging the growth of ground vegetation, scrub and young trees. This provides the okapi with a variety of foliage on which to browse. The animal's diet may also include seeds and fruit.

The okapi's senses of smell and hearing are good, but its sight is poor. It has scent glands on its feet and appears to use scent-marking to indicate favorite feeding areas. This suggests that okapis are territorial animals, a feature that corresponds with their generally solitary nature.

Encounters between rival males involve both real violence and ritualized combat. In ritual fighting, the animals shove each other with their shoulders—the hide is thicker at the contact points—and grapple with each other neck-to-neck. When one okapi threatens another, it lowers its head to present its horns, and if the fight gets serious, it may charge the other and deliver a head butt.

The female appears to take some of the initiative during courtship by uttering low, coughing sounds to invite the male, but clear display signals are needed from her partner before mating can take place. Females give birth to single calves during the rainy season, after a gestation period lasting 14-15 months. A mother okapi will defend her offspring by kicking with her feet, but many young okapis still fall prey to large forest-dwelling carnivores such as leopards.

The lofty giraffe

The giraffe is an animal of well-wooded savannah, where it browses in the treetops alongside grazing animals such as zebras, wildebeest and, in some areas, domestic cattle. It is able to reach the highest leaves, partly due to its long neck, but also because its legs are extremely long. Like most mammals, the giraffe has seven vertebral bones in its neck, but each is greatly

LEFT A female okapi in a European zoo. In the wild, okapi live deep in the rain forest of northern Zaire, where the undergrowth is thick and the foliage is within easy reach.
ABOVE RIGHT During a playful fight an adult okapi allows its offspring to play the role of a superior. The young animal hits the adult on the knees (A), asserts its authority by raising its head (B), and beats the adult who lies on the ground in a submissive position (C-D).
PAGE 425 A group of reticulated giraffes file across the African savannah.
BELOW A world map showing the distribution of giraffes and okapis.

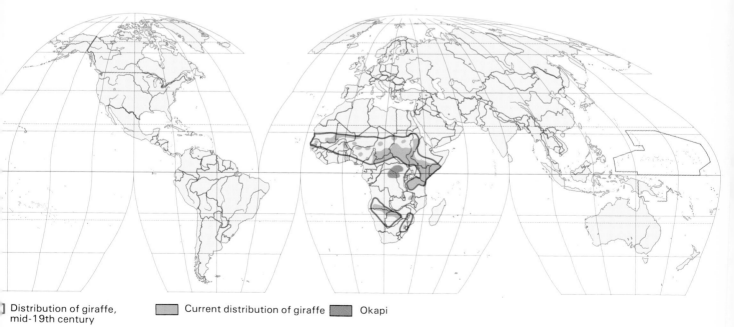

] Distribution of giraffe, mid-19th century   Current distribution of giraffe   Okapi

ABOVE At feeding time, an infant giraffe drinks its fill as its mother browses on a convenient tree. Giraffes are suckled for about 15-17 months. The young animals are left on their own for much of the day, while their mothers go off to forage. Without protection, they form groups for mutual safety. Their height (giraffes, even at birth, measure up to seven feet at the shoulder) enables them to spot approaching danger and make an escape. Even so, less than half the calves survive their first year.

elongated, and the neck is supported by powerful muscles. These muscles are anchored to dorsal extensions of the backbone that form a prominent hump behind the animal's shoulders. A fully grown bull giraffe may weigh over two tons, and reach over 15 feet in height.

Both male and female giraffes have horns, although the male's are thicker. The giraffe is one of the few ungulates actually born with horns. At birth, the horns are cartilage cores lying flat on the skull, but within a week they stand upright and start to become bony. Males also develop a number of lumps on their skulls as they grow older. A giraffe's back is short in comparison to its neck and legs, and because its hind legs are shorter than the forelegs, its back slopes down toward the rump. The animal's tail ends in a long tassel of black hair, which is useful for brushing away flies in the heat.

Though coat markings are not reliable guides to classification, the various subspecies of the giraffe differ in the general color and pattern of their hide. The coat of the reticulated giraffe consists of regular chestnut blocks divided by narrow white lines, while that of the Masai giraffe has irregular, jagged patches that may vary in color from dark brown to pale orange. Some 20 subspecies have been described, but a total of eight is now generally accepted.

Sharp-eyed lookouts

Because its head is held so high above the ground, the giraffe has an excellent field of view and, unlike the okapi, vision is its sharpest sense. Its ability to spot danger from afar is unrivaled, and many other grazing animals seek the company of giraffes since they know they can count on their keen sight to warn them of approaching predators. Giraffes have long, thick eyelashes that protect their eyes from injury when they are feeding on spiny vegetation.

Giraffes feed mainly on leaves, gathering them with their long tongues that can extend to 18 in. They often strip leaves from trees by drawing the twigs through their teeth.

Males and females generally feed at different levels among the trees. This is not only because the males are larger (they may be almost 3 ft. taller than females), but also because the two sexes have different feeding techniques. Males reach up to pull down foliage, while females tend to gather leaves that

ABOVE Two female giraffes quench their thirst at a waterhole. Taking a drink is an awkward business; although the giraffe's neck is very long, its front legs are even longer, and it has to splay them wide apart in order to reach the ground with its muzzle. It might seem easier for them to wade into the stream to drink, but giraffes are unable to swim and dislike entering water.

are growing beneath head level. The sexes are, therefore, not in competition for the available food. If necessary, they will also eat plants growing on the ground, although to do so—and to drink—they must splay their forelegs in an ungainly manner.

The lion, hyena and wild dog are the giraffe's main natural enemies, but the threat is mostly to the young. Only lions are able to bring adults to the ground, and they do so with difficulty. A solitary lion will often come off the worse in an encounter with a giraffe, and it is not unusual for it to be kicked to death by its intended victim. The giraffe's long front legs, armed with stout soup-plate-sized hooves, are formidable weapons. Like okapis, giraffes always walk with both legs on one side moving in unison, giving them a deceptive, ambling gait—in fact they can run at speeds of over 30 miles an per hour, when necessary. The main natural threats to full-grown giraffes are food scarcity and accidents involving falls and broken limbs.

Giraffes are sociable animals and band together into herds which are constantly altering in makeup. Each animal has a large home range surrounding a preferred central area, and will associate with any other giraffes who happen to be in the vicinity. Even mature males will accompany one another, provided they have established a rank order. At each encounter

ABOVE With its head held high and neck at full stretch — the characteristic browsing style of the males — a giraffe reaches for some leaves. By contrast, females browse with head bent downward and take leaves from lower down the tree. The different stances ensure that the two sexes do not compete for the same food, and means that the young can count on a supply of high-quality milk when their mothers return to the nursery group to suckle them.

between two males, status is reestablished by a display: the dominant animal raises his head with his neck upright, while the subordinate lowers his head submissively. If there is a dispute, the two will commence an action called "necking"—a ritualized fight that involves them standing next to one another and swaying their necks from side to side, entwining them in an attempt to outdo each other in a show of strength.

True fights will only occur if an entirely strange male, whose rank is not established, enters the area. In such cases the combatants will use their heads to hit one another. The pendulum-like movement of the neck causes the giraffe's head to hit an opponent with considerable force. The giraffe's brain is protected by a thick skull, well cushioned with fluid-filled spaces or sinuses, but despite such protection, fighting giraffes occasionally lose consciousness.

Although male giraffes are generally tolerant of each other, a few superior males dominate the mating in a particular area. Single young are born after a gestation of some 15 months, and they are already 6 ft. 6 in. tall at birth. The young are generally born in traditional maternity areas; a number of females may give birth at once, and after one or two weeks the young are left to join together in groups. These groups can fend for themselves, and this frees the mother giraffes to go off and forage. The mothers return to suckle the young at regular intervals, until they are weaned at about 15 months (males) or 17 months (females).

GIRAFFES CLASSIFICATION

The family Giraffidae consists of two species: the okapi and the giraffe.

The okapi, *Okapia johnstoni*, is a short-necked browser that lives in the rain forests of northern Zaire in equatorial Africa. The giraffe, *Giraffa camelopardalis*, has a much wider distribution through Africa south of the Sahara. It is found on grassland with trees and in open woodland. There are eight distinct subspecies, including the reticulated giraffe, *G. c. reticulata*; the Masai giraffe, *G. c. tippelskirchi*; Rothschild's giraffe, *G. c. rothschildi* (Kenya); and the South African giraffe, *B. c. giraffa*.

THE PRAIRIE SPRINTER

With a top speed of more than 45 miles an hour, the strikingly marked pronghorn can outrun most of its enemies on the open grasses and brushlands of North America

The North American pronghorn is an oddity among grazing animals—in some ways it resembles deer, but in others it is like antelopes. In the past it was classified in a family of its own. Today, however, it is considered to be part of the large and diverse family, the bovids, which includes the antelopes, the wild cattle and the gazelles.

The most remarkable feature of the pronghorn is its horns. They are hollow, like antelope horns, yet their outer sheaths are shed and regrown after the annual rut, like the antlers of deer (although in deer, it is the entire structure that is shed). Each projection consists of a black sheath on a bony base. New horns grow up from beneath, break open the old overlying horn and cause it to drop off. The horns are comparatively small, and rarely exceed 10 in. in length; those of the male are backwardly hooked, with forward-pointing prongs. Females generally possess simple, shorter horns without branches.

Pronghorns are long-legged animals with box-shaped bodies. They measure some 3 ft. at the shoulder, and a mature male can weigh up to 160 lbs.; females are slightly smaller. They have erect, pointed ears and large eyes with long black eyelashes that cut out the glare of the sun on the open prairie.

The animals have striking markings: the buck has a black facial mask and a black patch beneath each ear, and both sexes have brilliant white patches on the breast, flanks and rump that stand out against their tan coats. The effect can be enhanced at will, for pronghorns have powerful hair-erector muscles that allow them to bristle up the white patches to make them appear larger.

White flare

When a pronghorn is alarmed, it erects the hairs on its rump patch. This flares out like a white rosette, forming an alarm signal, clearly visible to other members of the herd at several hundred yards. At the same time, glands on the rump release strong alarm scents. It is possible for a human to smell these scents, so undoubtedly they are readily detected by other pronghorns, which have a much better sense of smell than we do. If danger is present, the whole herd takes flight, sprinting at speeds of 40 miles per hour or more until they have outdistanced their enemy (the record sprinting speed is 53.8 miles an hour).

LEFT Knee-deep in grass and brushwood, a buck pronghorn surveys his territory. Once widespread on the North American prairies, the pronghorn has been reduced to a number of scattered populations, and two of the four subspecies are endangered.
PAGE 431 A group of pronghorns, marked by striking white rump patches, pick their way over rough prairie grassland in search of succulent shrubs.

PRONGHORN CLASSIFICATION

The pronghorn, *Antilocapra americana*, is the sole member of the Antilocaprinae, a subfamily of the family Bovidae. It is found on the open prairies of western North America, from Canada down to Mexico, and there are four subspecies. Two of these, *A. a. peninsularis* and *A. a. sonoriensis*, are now classed as endangered species.

ABOVE **Young pronghorn twins follow their mother through the bush. They can outrun a man within a few days of birth but lack the stamina needed to keep up with the adults, so they spend the first three to four weeks hidden in dense cover. Female pronghorn** **frequently give birth to twins where food sources are plentiful.**
LEFT **The white hairs on the pronghorn's rump can be raised to act as an alarm signal, which is visible to the human eye at a distance of up to 2-3 miles.**

Pronghorns possess a complex array of scent glands, producing a range of different chemical signals. A male has nine scent glands in all, two on the rump, four between the toes, two beneath the ears and one above the tail. Females have six glands—two rump glands and four between the toes. The glands beneath the ears of the males are used in courtship and in marking out territories.

The social organization of pronghorns varies according to the time of year. The largest groups, consisting of up to a thousand males, females and young, are formed after the autumn rut. In early spring, the large groups break up, and females and young gather in herds of up to 50 animals.

The males establish territories and defend them against rivals from late March on, though mating does not take place until the rut. The rut lasts for a two- to three-week period, falling sometime between late August and early October, the precise timing depending on how far north the animals live. Territories vary in size from a minimum of 0.1 square mile to an area twenty times that figure, and males may return to the same location year after year.

Male pronghorns spend much time marking their territorial boundaries with scent. A black spot below the ear marks the site of the subauricular gland on the male. Scent secretions from this gland are used to mark tall grasses or large plants within the territory. When a receptive female enters a male's territory, the male approaches her, turning his head several times. A distinct waft of scent is blown toward the female by this action.

433

The pronghorn inhabits open grassland and brushland in western Canada, the western USA and northern Mexico.

The flehmen response

Like many other antelopes, male pronghorns display a curious behavior, known as "flehmen," as part of their courtship ritual. After sniffing or licking the female's urine, the male raises his head slightly and curls up his lips. Together with a pumping action of the tongue, this forces the scent from the urine to the Jacobson's organ located at the roof of the animal's mouth. Here the scent is sampled, and the information passed to the brain. The male can recognize the distinctive scent in the urine of a female in heat. When courtship is complete, the animals mate, with the male's chest resting on the back of the female, while his neck is held high.

Female pronghorns are unusual among mammals in that the number of egg cells (ova) they produce at one time greatly exceeds the number that, when fertilized, could develop into new offspring within the womb. From four to seven eggs are present when the female is ready for mating, yet she usually gives birth to only one or two young—if other eggs are fertilized, the resulting embryos die in the womb. Known as polyovulation, this is a reproductive strategy much more common among reptiles than mammals. Why it should occur in the pronghorn, along with the plains viscacha and a few other animals, is not fully known.

Too young to tangle

Females give birth to their young in spring, after a gestation period of about 250 days. They usually leave the herd to give birth alone. Twins are common in areas where the food supply is good. For the first three to four weeks of life, the young remain hidden in vegetation as their mother feeds nearby. She will return at intervals to suckle them—but, during this time, the fawns may have as little as half an hour's contact with their mother each day.

Females reach maturity at 16 months, but it usually takes much longer before males are able to breed successfully. A breeding male must be capable of defending a territory to which females will be attracted. A youngster is generally unable to hold a territory until he is three years old. Once he has reached that age, he should be able to defend a territory successfully for the next five years or so, before his strength starts to wane.

When the western states of North America were first colonized by Europeans in the mid-19th century, the pronghorn ranged all over the western prairies; there may have been as many as 40 million animals in all. Within 50 years, cattle rearing, cultivation and indiscriminate hunting had reduced their numbers to less than 20,000, and the species seemed to be slipping toward extinction. Fortunately the animals were placed under strict protection before it was too late. Conservation efforts stopped the slide, and today the population in Canada and the USA has risen to some 450,000, concentrated mainly in national parks and nature reserves scattered over a wide area. Now, approximately 40,000 pronghorns are allowed to be killed for sport each year.

In Mexico, the outlook for the pronghorn is not so encouraging. Only about 1200 pronghorn survive there, and their numbers are still declining due to the destruction of their habitat and illegal hunting. Numbers of one subspecies have dropped by more than 60 percent since the mid-1960s.

BOVIDS CLASSIFICATION

The Bovidae form by far the largest family within the even-toed ungulates, numbering over 120 species in all. The family is divided into six subfamilies, some of which are further divided into tribes.

The pronghorn is the only member of the subfamily Antilocaprinae, while the 17 species of duiker belong to the subfamily Cephalophinae. The subfamily Bovinae contains three tribes: the wild cattle, the Bovini; the four-horned antelopes, the Boselaphini; and the spiral-horned antelopes, the Strepsicerotini. The subfamily Hippotraginae also consists of three tribes: the reedbuck and its close relatives, the Reduncini; the gnus, hartebeest and their relatives, the Alcephalini; and the horse-like antelopes (oryx, sable antelope, etc.), the Hippotragini.

The subfamily Antilopinae comprises the dwarf antelopes of the tribe Neotragini and the gazelles of the tribe Antilopini. The goat antelopes form the subfamily Caprinae and are divided into four tribes: the Saigini (saiga and chiru); the Rupicaprini (chamois, mountain goat, etc.); the Ovibovini (musk ox and takin); and the Caprini (ibex, mouflon, etc.).

DOE-EYED AND DIMINUTIVE

The duikers and dwarf antelopes are
delicate, sometimes secretive creatures
and include one of the tiniest hoofed mammals,
the hare-sized royal antelope

Beira

Klipspringer

Oribi

Common bush duiker

Swayne's dik-dik

Yellow-backed duiker

The duikers are short-tailed, slender-legged antelopes with arched backs—the result of the hind legs being longer than the forelegs. The Afrikaans name "duiker" (pronounced die-ker) means "diver" and refers to the animal's habit of diving into the undergrowth when it is frightened or disturbed. The 17 species that make up the duiker subfamily are found in Africa, south of the Sahara, from Senegal in the west to Tanzania in the east.

Duikers range from the size of a hare to that of a fallow deer. One of the smallest species, the blue duiker, weighs only 9 to 13 lbs., while the yellow-backed duiker can weigh up to 180 lbs. Female duikers are often slightly longer than males. Both sexes have short, straight horns which point slightly backward and are sometimes concealed by a tuft of hair. The hooves are narrow and pointed. Their build and small size adapt them well to life in dense forests and thickets, through which they move easily in search of food. Being small also helps them hide from predators. Bush duikers, which live in more open habitat, are larger than some of the forest duikers but have a similar appearance.

A startling diet

Duikers lead active lives and need to eat large quantities of nutritious foods (especially fruits, leaves, flowers and buds of trees and bushes) each day to replace lost energy. Surprisingly for herbivorous mammals, they sometimes supplement this diet by catching and eating small nestlings and mammals. Duikers also eat some insects, snakes, eggs and carrion —animal food that provides valuable protein. They rarely eat grass, which is a low-energy food. Another advantage of the duiker's wide diet is that it contains enough water so that they seldom need to drink. The bushes, trees or herbaceous plants that duikers feed on are found in small patches and are able to support only the minimum number of animals. Duikers, therefore, live solitary lives or, more commonly, as mated pairs within territories which they defend. At most times there is enough food within each territory for a pair of duikers and their fawn.

Territories are marked with scent glands and defended. Both sexes scent-mark twigs and plants by rubbing them with clear or bluish secretions from the suborbital glands, which open below the eye. Marking is more frequent when there are other duikers in the

ABOVE Two blue duikers indulge in mutual marking; the animals rub scent onto one another from large glands near each eye. Duikers live in pairs within small territories which they defend from other duikers.

Both sexes are active in territorial marking and defense.
PAGE 435 The klipspringer is adapted to life on the rocks and grassy mountain slopes, and walks on the tips of its toenails, gripping the rock.

area. Although territories are small—seldom larger than 10 acres—they are rigorously defended. In duikers that form long-term pairs, such as Maxwell's and blue duikers, both sexes actively defend their territory. Generally, duikers are more aggressive to members of the same sex than to individuals of the opposite sex.

Young duikers may have similar coats to the adults (as in species such as the Zebra duiker and Maxwell's duiker), or their coats may be different. They are weaned at the age of about five months but do not reach sexual maturity until they are approximately one year old. The smaller species, such as blue duikers, mature earlier than the larger species, and females mature slightly earlier than males. When they are mature, young duikers must leave their parents' territories to find mates and establish territories of their own.

ABOVE The common bush duiker grooms itself on the hot plains. It is the only species of duiker living in savannah and open bush country.
BELOW The map shows the distribution of various duikers and dwarf antelopes.

Contrasting habitats

The common bush duiker resembles the forest duikers in many aspects of feeding and behavior. It differs chiefly by living in savannah grassland and open bush, in contrast to the dense forests preferred by its relatives. Bush duikers are more solitary than forest duikers, probably only forming pairs at mating times. Like other duikers, they have one young but commonly give birth twice a year.

Dwarf antelopes

The dwarf antelopes are found in Africa, south of the Sahara in a range of habitats as diverse as the animals themselves—from dense forests along the West African coasts (royal antelope) to dry rocky outcrops at up to 13,000 ft. in central, eastern and southern Africa (klipspringer). The term "dwarf antelope" is sometimes restricted to just three species: the royal antelope, the pygmy antelope and the suni.

Females do not usually have horns and are, on average, 10-20 percent larger than males. Dwarf antelopes range in weight from 3 to 5.5 lbs. in the royal antelope (which is 18-22 in. in length), to 30 to 60 lbs. in the beira (28-34 in. long). Coat color, markings, length of muzzle, length of legs, development of hooves and scent glands vary between species. Most

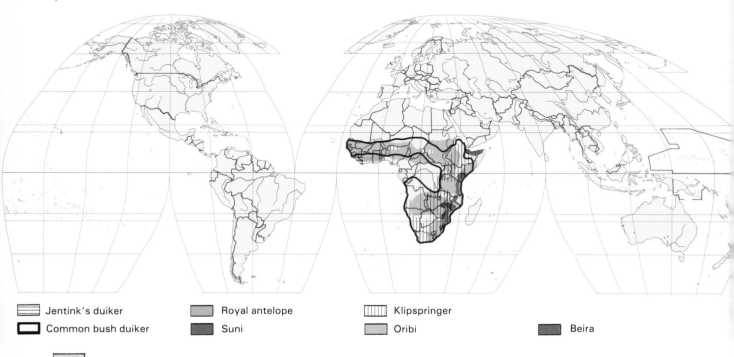

Jentink's duiker	Royal antelope	Klipspringer
Common bush duiker	Suni	Oribi
		Beira

dwarf antelopes are browsers, eating high-energy fruit, buds, leaves, roots and tubers, with the exception of the oribi, which feeds mainly on grasses. They live alone, in pairs or in small groups. Territories are defended by males and are scent-marked.

The three species of the smallest dwarf antelopes in the world live in forests and the dense undergrowth of woodlands in Africa. They are the royal antelope in West Africa, from Sierra Leone to Ghana; the pygmy antelope from southeastern Nigeria to western Uganda; and the suni from Kenya to eastern South Africa. Suni also live in bushy thickets in drier plains habitats at altitudes of up to 6500 ft. These tiny antelopes browse on fruit, buds and leaves in a manner similar to the duikers, but also eat grass and other vegetation.

A diminutive "king"

The royal antelope is one of the world's smallest hoofed mammals (only the lesser mouse deer is smaller). It is known locally as the "king of hares" because of its similar size and arched back. According to Liberian folklore, the royal antelope is renowned for its speed and cleverness, like the American folk character "Brer Rabbit." Like the pygmy antelope and unlike the suni, it is mainly nocturnal. When alarmed, these small antelopes escape by freezing and then jumping away at the last moment. Pygmy and royal antelopes can leap obstacles up to 8 ft. high, arching their backs as they do so and looking like small springbok. These escape tactics are important to the little antelopes, since their small size makes them vulnerable as prey to a wide range of predators, including snakes and birds of prey.

Very little is known of the social life of these secretive dwarf antelopes. They live mainly alone or in pairs, but sunis sometimes gather in small, loose groups at rich feeding sites. These groups never contain more than one adult male.

The dainty dik-diks

Dik-diks are small, graceful antelopes, slightly larger than a hare and with large ears and eyes and a short tail. They live in arid or semi-arid regions, on stony or sandy slopes interspersed with areas of scrub and bush. The three species occur in different parts of Africa. Gunther's dik-dik is found in East Africa, from northern Uganda eastward to Ethiopia and Somalia.

TOP Clearly visible on this common bush duiker is the erectile crest of hair on the crown of its head. The crest gives the duiker family its Latin name *Cephalophinae*, meaning "tuft-headed."
ABOVE The royal antelope is Africa's tiniest horned ungulate—some adults weigh less than four pounds. The animal's long hind legs and short front legs tilt its body forward so that it stands like a hare. In spite of its small size the royal antelope can leap over obstacles 8 ft. high.

439

ABOVE **A pair of Swayne's dik-diks. The animals mate for life and live in territories which they scent-mark and defend.**

Dik-diks mark the runways connecting the rock piles and bushes within the territory with secretions from their foot glands.

DUIKERS CLASSIFICATION

The 17 species of duiker belong to the subfamily Cephalophinae. They are all small to medium-sized antelopes and range over most of Africa south of the Sahara.

The forest duikers, of the genus *Cephalophus*, comprise all but one of the species. They are found in a variety of lowland and upland forest habitats. Members of the genus include Maxwell's duiker, *C. maxwelli*, of West Africa; the blue duiker, *C. monticola*, found in Central Africa, parts of Nigeria, and parts of East and southern Africa; the zebra, or banded, duiker, *C. zebra*, which is confined to a few forests in Liberia, Sierra Leone and the Ivory Coast; the yellow-backed duiker, *C. sylvicultor*, which has a wide range in Central Africa but a fragmented distribution in West Africa; and the very rare Jentink's duiker, *C. jentinki*, which survives in Liberia and possibly also in the Ivory Coast.

The other genus, *Sylvicapra*, contains only the common, or gray, bush duiker, *S. grimmia*, which inhabits savannah and open bush over most of sub-Saharan Africa.

Swayne's dik-dik occurs in northern and eastern Ethiopia and Somalia; while Kirk's dik-dik is the most widespread, with two populations—one ranging from southern Somalia to Tanzania and the other from Angola to Namibia. Gunther's and Kirk's dik-diks are found in denser cover than the several subspecies of Swayne's dik-dik and at altitudes reaching up to 10,000 ft. above sea level.

Like many dwarf antelopes, female dik-diks are larger than the males—weighing up to one-fifth more. Only the males have horns. Small and pointed, the horns are widely separated on the head and are angled slightly backward.

The hairs of the coat in both young and adults are speckled in a characteristic salt-and-pepper fashion. One of the most distinctive features of dik-diks is the small crest of hairs on the top of their heads which can be erected during threat displays. Dik-diks eat a variety of plants, fruit, leaves and seeds, but like other small antelopes, they do not eat much grass because of its low energy value.

An air-conditioned nose

The muzzles of Gunther's and Kirk's dik-diks are noticeably elongated. They have sometimes been described as "trunk-like" because of their length and because the upper part is slightly forked and can be moved. This strange development is part of the dik-diks' ingenious adaptation for remaining cool in the arid regions they inhabit.

Inside the long snout, blood in the veins passes near the surface of a nasal membrane. When the dik-dik breathes, the air in the nasal passage cools the warmer blood below the damp membrane. If it is very hot, or the dik-dik has been very active, the animal can cool off by evaporating heat from the membrane by "panting" through its nose, just as dogs pant to cool down after exercise.

Male and female dik-diks live in pairs, remaining faithful throughout their lives. Each pair will mark and defend a territory which contains piles of rocks (kopjes) and feeding bushes connected by a series of well-traveled runways. Territory sizes vary between

ABOVE Kirk's dik-dik is the most widespread species of dik-dik, being found in southwest as well as eastern Africa. It is a territorial animal, usually living in pairs. The young live with the parents until they are mature (at six to nine months).
PAGES 442-443 During a mating ritual, a female steenbok urinates in front of the horned male.

160 and 1650 ft. in radius, depending on factors such as the quality of the land and the population density of other dik-diks. Sometimes, larger groups may gather, but these are usually at feeding sites not "owned" by any particular pair.

Unlike many other small antelope, dik-diks have a well-developed sense of sight as well as a good sense of smell and hearing. Their small size makes them vulnerable as prey to a wide range of carnivorous mammals, birds of prey and snakes that share their African habitat; so having good senses with which to detect danger is an important means of avoiding capture. The alarm call of dik-diks is a sharp, ringing "dik-dik"—which is how they received their common name. Dik-diks are hunted locally by villagers for food.

Young dik-diks are born after a gestation of five to six months depending on the species. The young lie concealed for the first two to three weeks and are weaned by the age of three to six months.

The agile klipspringer

Klipspringers (in Afrikaans, literally "rock jumpers") measure 30-35 in. in length and weigh 22-33 lbs., two to three times as much as dik-diks. They are also taller, with the smallest klipspringer standing as tall as the largest (Kirk's) dik-dik. However, klipspringers lack the long muzzles and crown crests of dik-diks and are clearly distinguished by some features of their own. They are stocky animals with rounded backs and short, stumpy tails. Females are usually larger than males and do not have horns (except in one subspecies in East and Central Africa in which the females have weak horns).

Among the most agile of all the hoofed mammals, klipspringers are well adapted to life in the rocky habitats of Ethiopia, eastern Sudan, East Africa and coastal Angola. Their arched backs, similar in shape to those of the smaller forest-dwelling duikers, enable klipspringers to stand with all four legs bunched tightly together under their bodies. This allows them to perch on rocky outcrops, from which they can keep a sharp lookout for predators or rivals that might threaten their territory.

The most unusual feature of klipspringers is their long, widespread hooves. Only the very tips of the hooves touch the ground; in humans, this would be

ABOVE Kirk's dik-dik has an elongated snout in which heat is evaporated from blood vessels to keep the animal cool as it moves around its hot, dry habitat.

equivalent to a ballerina walking on the tips of her toes. The hoof tips are harder at their edges than in the center, which means that they wear down in rocky terrain to form a cup-like structure which helps them keep their grip. Another klipspringer specialization is its coat of hollow, brittle hairs which help protect the body from being damaged by frequent contact with hard outcrops of rock.

Klipspringers are small enough to be attractive prey to many predators. Living in fairly open habitats scattered with rocky outcrops and bushes, they have well-developed sight and hearing, like dik-diks. They are active by day and also on moonlit nights. During the day one member of a pair, usually the male, watches for predators from a rocky lookout while the other members of the group feed. If danger threatens, the sentry gives a loud, shrill whistle-like alarm call and the animals leap into the bushes for cover.

Steenbok and grysbok

The steenbok and grysbok are antelopes that belong to the same genus. Steenbok weigh 22-32 lbs. and measure 28-40 in. in length, including tail. The steenbok lives from Kenya south to Zambia, Mozambique, Angola and South Africa. Sharpe's

DWARF ANTELOPES CLASSIFICATION

The dwarf antelopes form the tribe Neotragini—part of the subfamily Antilopinae. Most of the 12 species are small animals, some weighing just a few pounds. They occur over most parts of Africa south of the Sahara.

The genus *Neotragus* comprises the royal antelope, *N. pygmaeus*, (the smallest horned ungulate), which inhabits dense forest in West Africa; the pygmy antelope, *N. batesi*, of Central Africa; and the suni, *N. moschatus*, which occurs in eastern Africa from Kenya to Mozambique. The dik-diks belong to the genus *Madoqua*. Kirk's dik-dik, *M. kirkii*, has separate ranges covering Tanzania and southern Kenya and Namibia and Angola. Swayne's dik-dik, *M. saltiana*, and Gunther's dik-dik, *M. guntheri*, both inhabit dry scrubland in Ethiopia and Somalia. Gunther's dik-dik also occurs in Kenya and Uganda.

The genus *Raphicerus* contains three species: the steenbok, *R. campestris*, which lives in lightly wooded plains in southern and East Africa; Sharpe's grysbok, *R. sharpei*, of Tanzania, Mozambique, Zambia and Zimbabwe; and the Cape grysbok, *R. melanotis*, which occurs only in the southern Cape. The remaining three species are the sole members of their genera. The klipspringer, *Oreotragus oreotragus*, inhabits rocky outcrops over much of East and southern Africa, along with two isolated areas in Nigeria and the Central African Republic. The oribi, *Ourebia ourebi*, occupies a large but patchy range spreading over most of sub-Saharan Africa. It inhabits grassy plains near water. The largest of the tribe is the beira, *Dorcatragus megalotis*, restricted to an area bordering the Red Sea and the Gulf of Aden in Ethiopia and Somalia.

ABOVE A female oribi rests while her mate keeps a lookout for danger. The longer neck and more upright posture of these open-country dwellers contrast with the arched back of the forest-dwelling duikers and pygmy antelope.

LEFT Oribis are the only true grazers among the dwarf antelopes.
BELOW (Left to right) A male oribi marks his territory by smearing secretions from a black subauricular gland beneath each ear onto twigs and bushes.

ABOVE A male oribi uses his ears to waft scent from the glands under his ears toward his partner during courtship. Oribis have more scent glands than any other dwarf antelope; they are placed at six different sites on the body.

Afrikaans. Grysbok lie low to avoid predators, springing up when their enemies are at a distance of 100-130 ft. and diving into the cover of bushes or holes, disappearing as if by magic.

The grazing oribi

Oribis are long-legged, slender antelopes which live on large grassy plains near sources of water. They can be found on rocky slopes up to 10,000 ft. in altitude. They have a huge but patchy distribution, occurring within a broad belt of Central Africa, from Senegal eastward to Ethiopia and Tanzania; they are also found in Zambia, Angola, Zaire and the eastern half of southern Africa. Overhunting and disease have combined to reduce their numbers in recent decades, and the animals have become extinct in many localities.

The back of the animal is slightly arched, as the hind legs are longer than the forelegs. Males have horns which are ringed at the base, rising steeply from the head and curving slightly forward.

Oribis are the only real grazers among the dwarf antelopes. They defend territories and live in pairs or small family groups. The young have a dark gray coat when born and do not develop the red, fawn and white coloration of their parents until they are five weeks old. When threatened, oribis are likely to run away from danger rather than attempting to dive for cover. The animal has a distinctive "stotting" or "pronking" gait, rather like that of a springbok, in which the antelope suddenly leaps vertically with all four feet off the ground, and then resumes fast running—an action that may confuse predators.

The beira

The beira is a very rare gazelle-like antelope with a narrow distribution along the borders of the Red Sea and the Gulf of Aden, in northern Somalia, and possibly also eastern Ethiopia. Standing taller than a klipspringer, and weighing up to 20 lbs. more, the beira is large for a dwarf antelope. Males have upright horns.

Little is known of the behavior of this inhabitant of barren, rocky hills and mountain slopes. Like the klipspringer, it is well adapted to life on stony ground, having short hooves with thick, rubbery pads. Beira eat leaves and herbs as well as some grass. They seem to occur mainly as mated pairs, but occasionally up to 12 animals, comprising two or three families, join together to feed.

grysbok ranges from Tanzania to South Africa; the Cape grysbok is restricted to the Cape Province of South Africa. All three species live in wooded habitats or on open plains near woodlands. They have narrow, pointed hooves and naked muzzles. The horns point upward and are smooth.

Grysbok are smaller and stockier than steenbok, weighing 15 to 30 lbs. and measuring 25-30 in., excluding tail. They have dark reddish brown coats, speckled with white on the back. The Cape grysbok is grayer in color and this probably gave the animal its name—"grys" meaning "gray" in

HORNS OF ELEGANCE

A variety of horns—from magnificent corkscrew
shapes to simple spikes—crown the
spiral-horned and four-horned antelopes
of Africa and India

Bushbuck

Nilgai

Nyala

Common eland

Sitatunga

Four-horned
antelope

Spiral-horned antelopes and four-horned antelopes are two tribes within the bovid family. The spiral-horned antelopes are among the most attractive of the world's antelopes and are found in various types of African scrubland. The animals vary greatly in the color of their coats, the number and position of facial and body markings, and the color of their crests and manes.

The greater kudu is one of the best known of all antelopes. It is found in East, Central and South Africa. Much larger than the lesser kudu, it has a longer crest of hair along its back and, in the male, a neck mane. Males weigh up to about 660 lbs., females up to 475 lbs. The horns have two-and-a-half spirals and two long keels— sharp ridges on the underside of the horns. The male's impressive horns are not fully developed until the animal is six to six-and-a-half years old. The hooves have a soft, inner glandular pad.

In *Green Hills of Africa,* the American writer Ernest Hemingway described his encounter with a greater kudu as follows: "Across the stream on the far bank at the edge of the trees was a large, gray animal,

BOTTOM A group of greater kudu drink at a waterhole. Greater kudu live in groups of up to 10 or occasionally more. They rest in thick bush during the heat of the day, only emerging in the late afternoon to drink and feed.
BELOW A male greater kudu rests his head on a female's neck during a ritualized stage in the courtship ceremony.
PAGE 447 A young male kudu (extreme right) playfully butts a fully grown male with his horns, which are still growing. It can take up to six-and-a-half years for the horns of the greater kudu to fully develop. The female kudu in the background is much smaller than the males and lacks horns.

white stripes showing on his flanks and huge horns curling back from his head…long-legged, a smooth gray with the white stripes and the great, curling, sweeping horns, brown as walnut meats, and ivory pointed, the big ears and the great, lovely heavy-maned neck, the white chevron between his eyes and the white of his muzzle.''

Greater kudus are social animals, commonly living in small groups of six to twelve consisting of mothers and their young. One or two males are sometimes loosely associated with these groups. Toward the end of the dry season, when food is scarce, troops of up to 100 animals may gather together.

The hunt for food

Although older males may defend territories in the breeding season, greater kudu are not basically territorial. They often travel widely in search of foliage and grass, making it impossible for males to defend static territories of sufficient size all year round. Greater kudus are selective feeders, nibbling the younger and more nutritious parts of leaves, plants and grasses. When they eat grass that grows in large stretches, it is possible for several animals to join together in a group and still find enough food to satisfy their needs.

Greater kudus are seasonal breeders, but the timing of the season varies with latitude. In more northerly populations, breeding takes place in October; further south this occurs in March. A male follows a receptive female, uttering a low "ummh" sound to indicate his interest and intentions. A single young is born after a seven-month gestation

Females mature when about one-and-a-half years old, males at up to two years. It is usual for males to leave their family groups when they are mature, often going on to form bachelor herds.

Striking differences

Of all antelopes, the lesser kudu shows the most extreme difference in coloration between the sexes. While the males are bluish-gray, females are reddish-brown. Both have 11-14 vertical white stripes on their coats. The male's dorsal crest extends forward as a short mane. Like the greater kudu, its horns have two-and-a-half spirals and two long keels, or ridges. The hooves are tough on the outside, with a soft inner glandular pad fringed by hairs. Males weigh 200-240 lbs., and females 130-155 lbs.

Lesser kudu live in pairs, or groups of up to six animals, inhabiting home ranges of up to one square mile. In their dense thornbush habitat, lesser

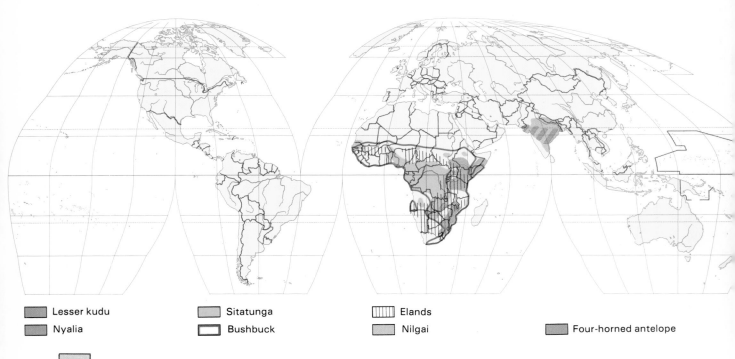

Lesser kudu	Sitatunga	Elands
Nyalia	Bushbuck	Nilgai

Four-horned antelope

kudu behave more like forest dwellers, concealing themselves to escape danger, although they will flee if it becomes necessary.

The nyalas

Nyalas live in riverside thickets in the savannah of southeast Africa. The animals are not as well known as kudu, but they have a similar appearance. Their coats are short and smooth: the male's coat is grayish-blue, with an ocher tinge; the female's is reddish-brown. Both have 11-13 vertical white stripes and three or four white thigh spots. The tail is longer and bushy in the male. Males have a mane of long hairs from their lower neck and shoulders, along the sides of the belly, to the thighs. The male's horns have one-and-a-half to two spirals with two long keels, and are lyre-shaped. Nyalas have good senses of smell and hearing, but poor eyesight.

The rare mountain nyala is different from its relative, living in high forests and heathland on a few mountains in Ethiopia. Its coat is shaggy; in the males it is dark grayish-chestnut, and in the females brownish-gray. There are two to five vertical white stripes and six to ten white spots on the flanks and thighs.

Discovered only in 1910, the mountain nyala soon came close to extinction, but protection has enabled numbers to recover to the present level of about 12,000. They are, however, still threatened by hunting, disturbance and habitat destruction. Both species browse on leaves, fruit and plants and do not graze.

ABOVE A pair of nyalas at a waterhole show the dramatic difference between the sexes, in both size and markings. The male is distinguished by his impressive horns and darker, fringed coat.
BELOW Attitudes of dominance or challenge in some bovids: the greater kudu (A), nilgai (B), black buck (C) and gaur (D).
FAR LEFT The map shows the geographical distribution of the four-horned antelopes and some of the spiral-horned antelopes.

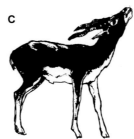

TOP The sitatunga is adapted to life in swamps. The long, splayed hooves spread the animal's weight as widely as possible so that it does not sink into the soggy ground. Females build raised platforms on which the young can lie safe from flooding. ABOVE The bushbuck prefers to live in areas of dense cover and wet environments, where it leads a chiefly solitary life, males and females occupying separate territories.

SPIRAL-HORNED ANTELOPES CLASSIFICATION

The spiral-horned antelopes are all native to Africa. They are medium-sized to large antelopes, the adult males reaching up to twice the weight of the females. The nine species make up the tribe Strepsicerotini—a group within the subfamily Bovinae.

Seven of the species belong to the genus *Tragelaphus*. The greater kudu, *T. strepsiceros*, inhabits woodland in East, Central and southern Africa, and the lesser kudu, *T. imberbis*, is found in areas of dense thornbush from East Africa north to the Sudan. The nyala, *T. angasi*, occurs in dense vegetation in southeastern Africa, while the mountain nyala, *T. buxtoni*, is restricted to highland parts of Ethiopia. The sitatunga, *T. spekei*, inhabits swamps and marshes along rivers in Central and southern Africa. The bushbuck, *T. scriptus*, is found throughout sub-Saharan Africa, in areas where vegetation cover is abundant. The bongo, *T. euryceros*, is a forest species found mainly in the lowland forests of Central Africa, though there are small populations in the uplands of Kenya and in West Africa.

The remaining two species belong to the genus *Taurotragus*. They are the common, or Cape, eland, *T. oryx*, and the giant eland, *T. derbianus*—the largest of all the antelopes. The common eland was once widespread in open habitats from southeast Sudan to South Africa, but is now largely restricted to national parks. The giant eland has declined drastically over much of its range running from Senegal east to Uganda.

Life in the swamps

The sitatunga lives in the swampy habitats of the Victoria Nile, Zambezi, Zaire and Okovango rivers. It is one of the few herbivores able to spend its entire life in swamps. Its shaggy, water-repellent coat is one adaptation to its habitat. Another is the splayed hooves which help to distribute its weight so it does not sink into soft mud or aquatic vegetation. Sitatungas are good swimmers and spend most of the day submerged in the water. One response to danger is to dive, remaining with only their nostrils above the surface, but at other times sitatungas take full advantage of their hooves and flee.

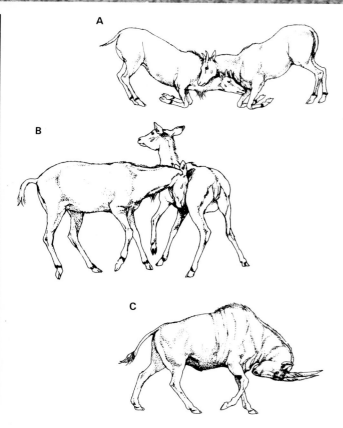

ABOVE Horns entangled, two male elands push each other back and forth as they test which is the stronger. These ritualized encounters make violent fights unnecessary.
LEFT The nilgai is not so restrained as the eland and engages in vigorous neck-to-neck combat (A). Even females will butt each other in the flanks (B). An eland (C) could easily kill its opponent if it used this butting technique.
PAGES 454-455 A group of female nyala drink from a waterhole in the Lengwe National Park, Malawi.

Male sitatungas sometimes fight during territorial disputes, kneeling on their forelegs and pushing forward and up toward the opponent, which must move to avoid the pointed horns.

Because of the marshy environment, sitatungas are at great risk when they are born. Many drown or are snatched by crocodiles. When their mother feeds, the young search out higher ground where they can rest. Some females trample vegetation to form a raised platform for their offspring. The young are weaned when four to six months old and become sexually mature at one to one-and-three-quarter years (females) and one-and-a-half to two years (males).

The sitatunga's tendency to form groups is weak, and many sitatungas forage on their own. The females are just as solitary as the males, and both behave aggressively toward others—clashes of horns are frequent, and

ABOVE Looking like a cross between a cow and an antelope, the nilgai or "blue bull" of India, probably resembles the ancestors of today's cattle, which developed from the antelopes only recently in evolutionary terms.
BELOW With such a short neck, the nilgai finds that it is able to graze more comfortably if it kneels down.

the animals bite and butt one another in the flanks with their heads outstretched. During courtship the male encourages the female to mate by maneuvering her into a subordinate position, forcing her hindquarters down by laying his chin across her back.

The tiny bushbuck

The bushbuck is one of the smallest of the spiral-horned antelopes, rarely weighing more than 150 lbs., although it is more bulky and robust in its proportions than many of its close relatives. It is also one of the most widely distributed, found throughout much of Africa south of the Sahara. It lives in areas where there is dense cover for it to hide in, and at altitudes of up to 13,000 ft. above sea level. The bushbuck's adaptability has enabled it to colonize a variety of habitats, and it swiftly moves into areas abandoned by other animals. It even adapts to plantations and other cultivated land.

Like the sitatunga, bushbuck tend to lead solitary lives. In order to find enough of the scarce, high-quality fruit, leaves and succulent roots which they eat, bushbuck forage singly or in small groups. Large herds of bushbuck would not find enough food in the same area and would have to spread out among the trees, perhaps losing contact with each other. The social structure that normally holds a large herd together therefore never developed as it did among the larger antelopes on the open grassland.

Bushbuck are unusual among spiral-horned antelopes in that they hold territories. Females occupy areas of about 35 acres, while male territories are twice this size. The areas are not strictly defended, and several individuals may share part of the territory. The jointly occupied section is usually the narrowest part of a roughly pear-shaped area. Despite tolerating these sharers, bushbuck try to avoid each other, and the mating season is marked by aggression between males and females in heat.

The elusive bongo

The bongo is the largest of the forest antelopes: a big male may weigh over 770 lbs., five times the weight of a bushbuck. The bongo lives in East, Central and West Africa, and is quite rare. Unlike other forest-dwelling antelopes, both the male and female have horns. In both sexes they may grow up to 3 ft. long in a smooth, open spiral twisting one to one-and-a-half turns.

The bongo has a reddish-brown coat with white, vertical stripes, and black-and-white markings on its legs, making it one of the most handsome of the antelopes. The markings also provide camouflage, since the fragmented pattern on its flanks breaks up the animal's outline in the broken light of the forest interior. It is active at dusk and during the night, and feeds on a variety of forest vegetation, including shoots, fruit and grasses, but mainly it browses on the leaves of trees and bushes. The bongo also digs for roots, using its horns, and may raid plantations of crops in some areas. Its deep-forest habitat has ensured its privacy, and little is known of its habits.

Cape eland

The common or Cape eland has been kept as a domesticated animal both in Africa and the Soviet Union. It is found from East Africa to the Cape Province in South Africa. Compared with other antelopes, it is a real giant, and a male may grow to 2000 lbs.; females weigh up to 1300 lbs. Such weights are comparable with those of small cattle, and in marginal country where cattle find difficulty getting enough food, the eland can be a more productive source of meat and milk. It is well suited to ranching, for unlike species such as the bushbuck and the bongo, it is an animal of open country and will often feed on grassland well away from cover.

The common eland is a sociable animal, forming herds that number several hundred in the breeding season. The herds wander from place to place in search of food; they may even move into arid country, such as the Kalahari Desert, to feed on scrub vegetation when seasonal rains encourage new growth. They are often found with zebra, roan antelope and oryx in the open country of southern Africa.

Resemblance to cattle

The general appearance of the eland emphasizes its close family ties with the cattle. Apart from its slim, deer-like head and long, spiral-fluted horns—present in both sexes—the eland is built like an athletic bullock, with muscular, squared-off haunches, deep flanks and a pronounced hump. The male's horns are on average twice the length of the female's and reach over 3 ft. in length. The animal has a tan hide with black-and-white markings on the legs. Eland in the northern parts of their range have faint, light-

ABOVE Tiny and delicate, the deer-like four-horned antelope is, surprisingly, a close relative of the larger nilgai. While the female, shown here, has no horns, males have two short horns between their ears and an even shorter pair between the brow ridges over their eyes. They are timid, solitary, nocturnal creatures, keeping within the cover of dense undergrowth in the woods of central India.

colored stripes on their forequarters. Despite their size, eland are quite agile. They feed mainly on grasses, leaves, young shoots and fruit, using their horns and hooves to break up the soil and pull up plants.

The giant eland (also known as Lord Derby's eland), a distinct species, used to range in a wide belt across sub-Saharan Africa from Senegal to southwest Sudan and northeast Uganda, but has been wiped out in most of its former habitats. Today only remnant populations survive. It is very similar to the common eland, but its coat is redder and has more sharply defined stripes on the body. A black collar rings the animal's neck and marks a fold of skin, or dewlap, hanging below the throat. The name giant eland is misleading, for it is not much larger than its relative, although its horns are longer (up to 4 ft. in the males). It is found in well-wooded country where it eats tree and bush foliage, and only occasionally ventures out onto the open grasslands.

457

THE LESSER KUDU
—A GRACEFUL AFRICAN ANTELOPE—

Of all the African spiral-horned antelopes, the two species of kudu are perhaps the most handsome. They are slim and well proportioned, with elegant horns twisting up in a loose, flamboyant spiral and gray or tawny coats etched with fine vertical white stripes.

The lesser kudu is not as imposing as its larger cousin, the greater kudu, but it is no less attractive. Like most other antelopes, it is a woodland animal, but it also finds food in dense grassland scrub where it can feed in the security of bushes and small trees. Although agile, kudus lack the sheer speed and stamina necessary to outrun a predator in a straight chase, so they normally avoid feeding out on the open plain. In any case, they prefer a more varied, high-quality diet than the grassland can provide on its own. They like to pick and choose among the grasses, bushes and trees, selecting the most nutritious food such as fruit, succulent leaves, tubers—which they dig up with their horns—and flowers.

One result of this diet is that kudu cannot live in large herds. Most herd animals are grazers—like domestic cattle—whose diet consists largely of grass and other abundant but nutritionally poor plants. In the case of the lesser kudu, competition for the choice of food they prefer would be too great if they foraged in herds, so they move around in small groups.

Kudu groupings

Many kudu males live solitary lives, especially when they get older. Otherwise, lesser kudu live in four types of groupings: male groups, female groups, harems and mixed groups. The male groups are usually very small, consisting of two to five adults and subadults. Groups of females and their offspring are larger, with eight or nine individuals. These are generally blood relatives. A harem consists of two or three females and their young, dominated by one adult male. The mixed groups are the most numerous, containing up to 20 lesser kudu, though most groups number between six and eleven. But even the larger mixed groups are small when compared to the herds formed by plains grazers like the wildebeest and zebra.

The average home range of lesser kudu individuals or groups is about one square mile—roughly what one would expect in an antelope of this size. Subadult males, like other herbivores, have a greater tendency to roam, and therefore occupy more space than mature males. Females occupy an area of about 0.7 sq. mi. The size of the range varies according to the season and the animals' grazing needs. In dry months, for example, the range may be extended so that the kudu have a larger area in which to find food. Kudu do not regard their territory as exclusive and make no attempt to defend it.

The social organization of kudu is flexible. Females associate with their relatives, forming the nucleus of the group, which is enlarged at intervals by newborn young. Other animals come and go, but their membership in the group is less secure. Females tend to be gregarious, and fewer than 10 percent

stay solitary. Young males usually leave their mothers long before they mature, at around a year or 18 months old (some leave as soon as they are weaned, at only 14 weeks) and wander on their own or join small bands of bachelor males of their own age. On rare occasions a male may remain with the mother, but his behavior toward her has a sexual element in it.

Born in the rain

Mating takes place at any time during the year, although there are peaks of sexual activity. These are linked to changes in the environmental conditions. An increase in food supply encourages mating. In some areas, mating is timed to ensure that the calves are born during the rainy season, when new vegetation provides the mother with the nourishment she needs to suckle her young.

When courting, the male kudu raises his head and lifts his chin slightly. The female responds by gradually raising her head and neck. She may deliver a series of hard blows to his flanks, indicating she has little intention of cooperating. But when the female raises her head and lets the male lick and nuzzle her throat and chin, the ritual courtship approach, and mating follows.

FAR LEFT The graceful lesser kudu prefers to forage in areas where trees and scrub provide cover and a variety of plant material to eat.
LEFT With its striking markings and majestic spiral horns, the lesser kudu buck is one of the most attractive of the great African antelopes.
TOP A female lesser kudu, startled by a human intruder, gets ready to bolt for cover.
ABOVE Two stages in the courtship of the greater kudu: (top) the male display of superiority, and (bottom) the female reaction, nuzzling her head against the male's flank.

459

A battery of threats

Female eland live in small bands that gather together into larger herds of up to several hundred (unusual for spiral-horned antelopes) during the mating season. These groups are often accompanied by one or more males, but the association seems to be casual. Most males live in bachelor herds, although older males tend to be solitary. They establish a hierarchy by regular confrontations, using a variety of threat signals that include head-lowering, horn-pointing and, finally, horn-tangling.

At the outset of this last ritual, the eland thresh aromatic shrubs with their horns, covering them with strong-smelling, sticky sap. They then force their horns into mud—often where another eland has urinated. Finally, the eland engage in a pushing contest with horns interlocked. During the mating season, it is the dominant male that sires most of the calves.

Encounters to establish dominance are highly ritualized, and the rivals rarely damage each other. They have strong necks and tough hide on their flanks to resist both the twisting effect of the tangling bout and the occasional thrust of an opponent's sharp horns. These are often used as defensive weapons; female eland are quite prepared to stand their ground and kill small predators in defense of their calves, and they are even capable of severely wounding large predators such as lions. They also attack by kicking with their front feet, an unusual habit among horned ungulates and one that is more often seen among hornless species such as horses.

Ancestral survivors

Although the four-horned antelope looks like a roe deer, it is a survivor of the primitive tribe of ungulates that gave rise to the cattle. Only one other member of the group still exists: the horse-like nilgai. Both species live in the woods of central India.

The most distinctive feature of the four-horned antelope is its second set of horns. They are modest structures: the rear pair rarely exceed 4 in. long, while the front pair, which grow just above and between the animal's eyes, are from one to two inches long. The female has no horns at all. The four-horned antelope grows to weigh some 45 lbs., and is delicately built with long, slender legs, a well-proportioned body and fine features. A forest animal, it lives in shady, damp woodland and is most

abundant in hilly areas. It is an elusive creature and has yet to be studied in depth. The animals live alone or in pairs and may be territorial. One or two offspring are born following a gestation of approximately seven and a half months.

A blue bull

The nilgai is a clumsy-looking animal with a massive body and very long legs. Its withers are higher than its hindquarters, giving it a curiously humped appearance. A large male may weigh nearly 600 lbs. and has two horns that grow to 8 in. long; the females are hornless. It is found to the south of the Ganges throughout the Indian peninsula; the arid bushland of Pakistan is the western limit of its distribution. The name nilgai means "blue bull" in Hindi and refers to the iron-gray coat of the mature male; females and immature males have tawny coats.

Nilgai are gregarious animals. The females live in herds, while adult males gather in bachelor herds of up to 20 individuals. Some males are also solitary. Local conditions determine the mating season, which does not conform to any set time. In some regions, the females may go into heat at any time of the year, while in other areas, such as east Rajasthan, mating is restricted to November and December. The males gather harems of two to ten females and defend them against rivals, continually rounding them up to ensure that they do not stray beyond the boundaries of the territory.

One or two offspring are born after a gestation period of eight to nine months. A female with newborn young tends to go off on her own, or with a female companion who has offspring of her own. When the young nilgai reach the age of a year or so, their mothers become more sociable, and the herds re-form.

STALWART GRAZERS

Long valued by humans for their hardiness,
the wild cattle are adapted to a range
of environments, from tropical swamps
to freezing mountain slopes

Cape buffalo

Gayal

American bison

Lowland anoa

Yak

Banteng

462

Cattle are among the most familiar of the ungulates, used for their milk and meat, and as draft animals. European domestic cattle are considered to be descendants of the aurochs, a wild form resembling Scottish Highland cattle that died out early in the 17th century. Elsewhere in the world, other wild cattle have been domesticated—notably the Asian water buffalo, which has been introduced in North and East Africa, southern Europe, South America and Australia. The wild yak manages to survive in the bitterly cold deserts and alpine tundra of the Tibetan plateau north of the Himalayas. The domesticated form of the yak is the traditional beast of burden in Tibet, as well as a source of meat and milk.

Cattle are highly adapted to life as grazers and browsers, with very efficient ruminant digestive systems, bulky bodies and broad heads with widely spreading horns in both sexes. The animals developed in the Old World, quite recently in evolutionary terms, from ancestors that had a great deal in common with present-day four-horned antelopes like the nilgai. The earliest types of cattle were adapted for life on the plains in the warm conditions of the Pliocene epoch (two to seven million years ago). As the climate deteriorated, some Eurasian species in the late Pliocene (two to three million years ago) moved south into Africa to escape the cooling climate. Other forms appeared that could tolerate cold weather.

Such animals were well equipped to survive the ice ages of the Pleistocene epoch (two million to 10,000 years ago), and during this period the ancestors of the American bison migrated across a land bridge, where the Bering Strait is today, into North America. Bison, yak and the descendants of the aurochs represent the cold-climate types, while the Asiatic and African buffaloes exemplify the warm-climate species.

Island anoas

The Asiatic buffalo genus includes the anoas, primitive species that preserve many antelope-like characteristics and could be regarded as a link between the four-horned antelopes and true cattle. Anoas are the smallest of the cattle, rarely growing to 650 lbs. in weight; there are two species, both restricted to the island of Sulawesi in Indonesia. The lowland anoa lives in undisturbed wet, lowland forest, while the mountain or highland anoa is a smaller type found in upland forests.

BOTTOM Two water buffalo enjoy a good wallow. The mud keeps the animals cool and helps get rid of skin parasites such as ticks, which become cemented into the mud as it dries and fall away when it flakes off.
BELOW In this encounter between two male gayals, the subordinate one (left) points its muzzle and neck to the ground, while the other demonstrates his higher status.

PAGE 461 Bison graze in Custer State Park, South Dakota, USA. Once there were millions of bison in North America; today only scattered populations, totaling some 50,000 animals, remain.
PAGES 464-465 A waterhole provides a welcome drink for a herd of Cape buffalo on the parched African plain. Herds may contain from as few as 50 animals to over 1000.

ABOVE A Cape buffalo pauses to scratch its head. The massive horns of these animals help to cushion their skulls during head-to-head fights and **also provide effective defense against predators such as lions.**
OPPOSITE The map shows the world distribution of some of the wild cattle.

The body of an anoa is rounded, but its legs are slender. Its horns point backward and are hardly curved at all, giving the animal a less bovine appearance than most cattle. It is also not as sociable as other cattle, tending to live alone or in pairs rather than in large herds. Hunting and the reclamation of their swampy, woodland habitat have reduced the numbers of anoa.

An isolated species

The tamarau (or tamaraw) is a similar species to the anoa, both in appearance and status. It, too, lives in the forests on an isolated island—Mindoro in the Philippines—and is in danger of extinction. Little is known of its habits, although it appears to be most active at night. Local people consider it a dangerous animal. The tamarau's survival depends on the conservation of its few remaining habitats and the uniting of the fragmented population. By the mid-1970s, the total population was estimated at less than 300. Today it is probably under 200. The tamarau lives in several scattered groups that are unable to meet up and interbreed; as a result, each group is likely to suffer from inbreeding, to the detriment of the species as a whole.

The Asiatic water buffalo is among the largest of the cattle. A male may stand over 6 ft. at the shoulder and weigh over a ton, and a well-grown female is not much smaller. Their long, sickle-shaped horns point backward and are flattened on top. In ancient times, water buffalo were found not only in Asia (including Mesopotamia) but also in North Africa and probably Europe too. Wild water buffalo are now restricted to Asia, but domesticated buffalo have been reintroduced in Africa and southern Europe and as far afield as South America and Australia, where the population is feral. There is also a small feral population in northern Tunisia. There are several subspecies of water buffalo, although some of these have become extinct in the wild and survive only under domestication.

Mud bath

The name "water buffalo" refers to the animal's habit of spending a large part of each day up to its shoulders in a pond, lake, swamp or mud wallow. The water or mud helps it to regulate its body heat, and the coating of mud on its skin forms a barrier against both the sun and parasitic insects. The buffalo eat water plants if they are available, as well as the usual bovine diet of grasses, leaves and young shoots.

Wild water buffalo are organized into herds of up to 30 individuals, made up of females and their young, sometimes with several young adult males. Older adult males live alone or in small bachelor herds of 10-15 animals each. Among the feral water buffalo of Australia, which live on the open plains, these small herds gather together to form large herds of 300 or more. In Asia the buffalo herds coexist well with other ungulates, including the rhinoceros, but there is some rivalry between buffalo and elephants; buffalo attacks on domestic elephants have been reported.

The domestication of water buffalo began thousands of years ago, and for several centuries they were the most important type of livestock in ancient Egypt. The animals were introduced in Italy in about AD 600, and some 100,000 still exist there. They later became established along the Danube and in southeastern Europe, where they remain fairly widespread. Throughout the world, an estimated 75 million domestic buffalo survive, although in some places, such as northern Australia, they have reverted to the wild and live as feral herds.

The placid, docile nature of buffalo, coupled with their considerable weight and power, makes them ideal draft animals, either for agricultural work such as plowing or for use as beasts of burden. They are particularly suited to working heavy, wet land in humid climates, since this is their natural habitat and they are resistant to most of the diseases endemic to such areas. They are also easier to look after than other draft animals; they will live on almost any type of forage, including unpromising material such as rushes and sedges, and they can be kept in the open, since they are able to stand up to cold weather. The only essential requirement is plenty of water.

The female water buffalo is an excellent source of good-quality milk, yielding over 500 gallons per year. A well-fed female will remain fertile for 17 years and may produce calves throughout that period at intervals of 14-15 months.

Wild and domestic populations often interbreed in areas where wild buffalo still exist, or where formerly domestic animals have returned to the wild. For this reason, and because there has been very little of the selective breeding that has transformed European cattle, wild and domestic buffalo are similar in appearance, with slate-black hides and white legs. In areas where there are no wild buffalo, domestic animals tend to be smaller, with shorter horns.

The African buffalo

The African buffalo occurs as two subspecies: the Cape buffalo and the forest buffalo. The Cape or black buffalo is a massive, heavily built animal, which measures up to 6 ft. high at the shoulder and may weigh 2000 lbs. It has a dark gray or brownish-black coat. The forest or red buffalo is significantly smaller, growing to 4 ft. high and 770 lbs. in weight. It is reddish-brown in color, and its horns are much shorter.

Between them, the two subspecies of the African buffalo are able to colonize a wide range of environments, including rain forests, open woodlands and savannah grasslands. They vary their way of life according to their environment. In common with many forest-dwelling ungulates, the forest buffalo is an unsociable animal. Some live in pairs or small groups, but many are solitary. The Cape buffalo, on the other hand, prefers more open terrain, where it forms large herds of from 50 to 1000 or more animals. On the

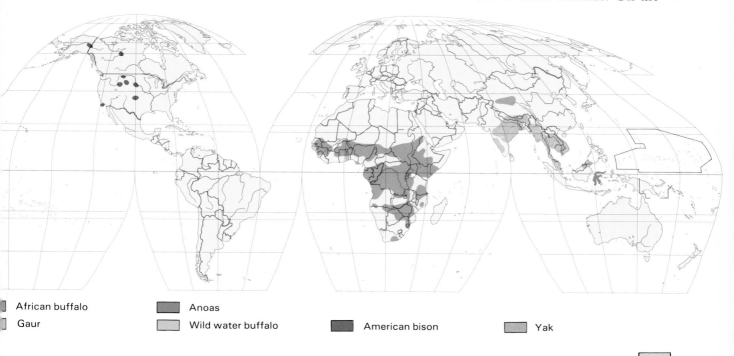

| African buffalo | Anoas | American bison | Yak |
| Gaur | Wild water buffalo | | |

WILD CATTLE CLASSIFICATION

The 12 species of wild cattle are all members of the tribe Bovini, a subgroup of the subfamily Bovinae. There are four genera. The Asiatic buffaloes *Bubalus* comprise four species: the water buffalo, *B. arnee*, which is now rare in the wild over much of its range in India and Southeast Asia, but is widespread as a domestic animal; the very rare lowland anoa, *B. depressicornis*, and the mountain or highland anoa, *B. quarlesi*, both found on the island of Sulawesi in Indonesia; and the equally rare tamarau, or tamaraw, *B. mindorensis*, which lives in forests and swamps on the island of Mindoro in the Philippines.

The genus *Synceros* contains only the African buffalo, *S. caffer*. Two subspecies are normally recognized: the Cape buffalo, found in open woodland and savannah, and the forest buffalo, which lives in the dense equatorial forests of Central Africa.

True cattle

The true cattle of the genus *Bos* comprise five species. The familiar farmyard cattle are generally considered to be the same species as their ancestor the aurochs, or urus, *B. primigenius*. The aurochs is now extinct, but "wild" herds of domestic cattle occur (notably the wild white cattle associated with the Chillingham estate in England). The banteng or tsaine, *B. javanicus*, is a similar animal to the farmyard cattle, found in the forests of Southeast Asia. The gaur, Indian bison or seladang, *B. gaurus*, is another woodland species, inhabiting upland forests in India and Southeast Asia. The yak, *B. mutus*, is found in the cold deserts and tundra of Tibet, at heights of 13,000-20,000 ft. The other member of the genus is the kouprey, *B. sauveli*, an extremely rare inhabitant of the Kampuchean forests.

Bison

There are two species within the genus *Bison*. The European bison or wisent, *Bison bonasus*, became extinct in the wild in 1919 but has been reintroduced in the Bialowieza forest in Poland, and in several sites in the USSR. The American bison, *B. bison*, was formerly widespread on the prairies and in the forests of North America, but was almost exterminated in the 19th century. Today it is found only in nature reserves and national parks. There are two subspecies—the plains bison, *B. b. bison*, and the more northerly wood bison, *B. b. athabascae*.

open grasslands of the Serengeti National Park in Tanzania, the average size of the buffalo herds works out at some 350 individuals.

Among the Cape buffalo, the largest herds are those composed mainly of females and young under two years of age. Some older males live alone, and others live in single-sex groups that may contain up to 20 animals, but more usually number three or four. The remaining males spend much of the year mingling with the female herds.

Within the buffalo herd, young females aged up to two years tend to stand near their mothers. Young males of the same age are not so attached to their parent, but they still remain in her vicinity. Subadult males of three to four years tend to join together within the herd; they avoid contact with adult males but will often stand with adult females.

During the dry season, males from five to ten years old tend to leave the main herd to form small bachelor herds, often grazing with the old males that are past breeding age. They return to the females during the rainy season and leave the old males to their own devices. The size of the mixed herds and single-sex herds, therefore, varies with the seasons.

Staying put

Elderly males over 10 years old normally leave the large herds for good, although they will graze alongside the younger animals if it suits them. They tend to take up residence in a particular place and keep to that spot. They are not strictly territorial, but simply more settled than the herd males, which may roam several dozen miles in a day within the same kind of terrain. By contrast, in three years a

ABOVE Disturbed while grazing, a group of Cape buffalo regard the intruder with suspicion. Buffalo are very dangerous to man, for they have a strong mutual defense instinct. If one member of the group is startled or injured, the others will come to its aid and may drive off or kill an enemy. Each buffalo recognizes the members of its own family group, and although one group may join other buffalo groups to form large herds, they never lose their own group identity.

group of elderly "settled" buffalo may move no more than half a mile from the center of their chosen patch. Their sedentary way of life offers undoubted advantages as a means of saving energy, but away from the herd the elderly males are at considerable disadvantage in terms of safety: isolated buffalo are much more vulnerable to attack. A large, compact herd of buffalo, well armed with horns, presents a formidable problem to a predator, and an animal within the herd is usually quite safe. There are reports of lame and even blind animals surviving for several years in the security of the herd. Since the buffalo are able to stand their ground in defensive formations, the herds also provide effective protection for newborn young. Infant calves would be unable to keep up if the group took to its heels when predators approached.

In the interests of mutual protection, the buffalo have developed an instinct for coordinated behavior: they all rest together and graze together, changing from one activity to the other within a few minutes. Such coordination helps to ensure that an individual buffalo is not left in an isolated position, making a tempting target for predators such as lions or hyenas. The success of this strategy is apparent from the observation that most buffalo killed by lions turn out to be lone and elderly males.

469

ABOVE **A herd of gaur, grazing peacefully in a woodland glade, typifies the way of life of most wild cattle. Unfortunately, their domestic descendants share their diet and habits, and animals like the gaur have become rare; their** habitats have been taken over to provide grazing for herds of domestic cattle. They are also affected by the diseases that afflict their domesticated cousins, and cattle plague (or rinderpest) takes an annual toll of wild Indian gaur.

Rallying to the cry

If one animal is in trouble, the whole herd will rally around in response to its alarm call. In one experiment, a recording of an abandoned calf's cry attracted a herd of 300-400 individuals to within a few yards on each of the six occasions it was played. The whole herd took part in the search for the calf, and each adult used calls similar to those which are uttered by a mother to her own offspring.

Although the whole herd may act together in many circumstances, buffalo first and foremost belong to small groups linked by family ties, the members of which tend to be particularly protective toward their relatives. The animals can recognize members of their own family group, and although they may mingle on the grazing lands, the buffalo usually sort themselves out into their respective groups at regular intervals.

Buffalo society is organized in hierarchical fashion, with rank descending in a long chain through the herd. The dominant male is superior to all the other males, the second in rank is superior to all but the first male, the third to all but the two above him, and so on. The system exists among females as well as males. In a very large herd, however, one chain does not run through the whole group. In such cases the hierarchy operates on the basis of subunits, so that a large herd may contain a number of dominant animals of equal rank, each a leader of its own tribe.

Rank is established by customary displays of aggression, threat and submission. During aggressive encounters, the two animals circle one another, sometimes nose to tail, with their shoulders hunched, necks slightly raised and muzzles pointing to the ground. The higher-ranking animal will often threaten the other with a flourish of his horns, sometimes striking trees and bushes. If this display proves insufficient, the two rivals face one another and fight head-to-head.

Avoiding contact

The huge horns of African buffalo meet in the middle at their roots to form a thick protective pad over the cranium. Even so, a fight can lead to serious injuries, and the animals will normally try to avoid real combat if they can. One way to do so is for the lower-ranking male to express submission before the fight begins, by holding his head in a horizontal position below the shoulder line, approaching the dominant animal, and then moving his muzzle close to the neck, hump or back legs of the animal, much as if he were suckling. This corresponds with observations of many species that suggest there may be a link between submission and infantile behavior.

Sometimes a threatened animal may adopt a posture of alarm or readiness for flight: the animal retreats a few yards, then stops and lowers his back while holding his head high with his nose pointing upward. In most cases this action defuses the situation, and the aggression of the dominant animal disappears.

Unlike other wild cattle, the African buffalo does not paw the ground or bellow before an attack; instead, it digs its horns into the earth or the soft mud at the edges of waterholes. Mud-covered horns are therefore a sign of aggression, since they are always exhibited before fights between males.

Members of the wild cattle tribe are native to North America, Africa, Europe and southern Asia.

Mock battles occur throughout the year, but real aggression is reserved for the mating season, when the males fight to gain access to females. Mating does not always occur at the same time every year, since it is greatly influenced by environmental conditions; the females go into heat at the beginning of the rainy season, and if the season is late they remain infertile. The young are born after a 300 to 340-day gestation, just before the following rainy season in a normal year; this ensures that the mother is well supplied with fresh vegetation to maintain the quantity and quality of her milk. Young are weaned at about six months.

A forest giant

The gaur of India and Southeast Asia is the largest of the true cattle and a near relative of the domestic cow. It is an impressive beast: a bull may grow to nearly a ton in weight, with a massive, shiny, blackish-brown or reddish-brown body, white "stockings" on its legs, a big head with long, upswept horns, and a prominent hump over the shoulders formed by extensions of the spinal bones.

Like many other wild cattle, the gaur has suffered the twofold pressures of land reclamation for agriculture and competition from domestic cattle. Originally it lived throughout the forested regions of India and Southeast Asia, but the fragmentation of its habitat inevitably led to the reduction and dispersal of the gaur population. Today it occupies three distinct areas in India: parts of the south, the central highlands of Madhya Pradesh, and an area along the slopes of the Himalayas above the Brahmaputra Basin. We know very little about its distribution further east. It appears to be present only in northern Burma, the national parks of Khao Yai and Tung Slang Luang in Thailand, and the Taman Negara National Park on the Malay Peninsula.

The gaur prefers extensive forestland with a plentiful supply of green plants, woody plants (especially bamboo), shrubs and saplings. It is normally found in quite hilly areas with reliable water supplies, at altitudes reaching up to 6000 ft. above sea level.

For much of the year many male gaur live alone or in single-sex groups. Adult males are usually solitary (88 percent of lone adults are over five years old), while subadults form small groups containing two or three individuals.

ABOVE A big bull gaur (right) urges a cow and two young calves onward. With their large horns, glossy coats, massive shoulder humps and deep dewlaps, gaur are among the most impressive of the wild cattle. The males assert their superiority over others by demonstrations of physical bulk rather than by fighting. Despite their strength, many adult gaur fall victim to tigers, although they may join forces to fend off an attack.

The small bachelor groups are not stable associations, although two young males may sometimes form what appears to be a steady friendship. Other males are found in herds containing females and young, and during the mating season most males will temporarily join these herds.

Herds may contain over 40 animals, with numbers and composition varying through the year. At any one time outside the breeding season, a typical herd might comprise one adult male aged over five, seven adult females and young, one subadult male, four males about one year old, and eight females of the same age. The herds are fairly stable, in that the same structure may remain for several weeks, but during the mating season the animals come and go quite frequently, and an adult male rarely stays with one herd for more than two days.

One gaur threatens another by making a bold approach with head slightly lowered, accompanied by a series of up-and-down jerks and side blows with the horns. A hit with the tip of a horn is sufficient to make most rivals retreat. Occasionally one threatening animal will approach another, snorting and shaking its head, sometimes striking its horns against saplings or bushes to make its intentions clear. In such cases the subordinate animal may turn and move away, or point his muzzle and neck to the ground in submission.

ABOVE At first glance, the banteng of Southeast Asia look like domestic cattle, and indeed many of these animals have been domesticated. Some have been crossed with other cattle such as the Asiatic zebu. Since wild banteng often interbreed freely with domestic banteng, hybrid characteristics may pass into the wild population.

The most impressive form of display among gaur is that known as lateral imposition, when each male seeks to overawe its rival by displaying the size of its dorsal hump. The two animals face one another with heads lowered and backs hunched, then circle around with stiff-legged steps. Eventually the initiator turns its flank to display the hump, which is made more conspicuous by the tensing of the back muscles. In establishing rank, it is the relative size of the animals that counts; physical strength and horn length are of little importance. Such a simple yardstick is useful during encounters with strangers, when a convenient system for assessing each other's status is required to prevent potentially damaging fights from breaking out. Most aggressive encounters take place between sub-adults—in both males and females. The animals push each other with their foreheads together, twisting their heads and varying the angle of thrust by moving the position of their bodies.

The song of the gaur

The mating season for the gaur varies according to location and may occur at any time between November and April. The males announce their readiness to mate by roaming through the forest, emitting a strange series of calls that can be heard well over a half-mile away. The calls are based on a clear "i-i-i-i" cry that lasts up to three seconds. The intensity may vary, or it may stay the same throughout the call. The first burst of notes is followed by another that is lower in tone and then a third even lower, the series running like a musical scale. Gaur utter their song in a variety of circumstances: when the animal is on its own, when it approaches a herd, when it is near a female, or when responding to the call of another male.

If a female gaur hear a male's song, she often moves toward the source of the sound. When a male encounters a herd of females, he may circle the animals several times or go in among them to assess which are ready to mate by sniffing at their genital areas. If the male finds a female in season, he exhibits the "flehmen" response: throwing his head back and curling his top lip. He then stays near the female, watches her, and from time to time advances toward her. At first she retreats, but when she is ready to mate she will stand still and allow him to mount. The pair usually remain in or near the herd during mating and stay close to one another for the rest of the day, exchanging licks to the neck, head, shoulders and back.

A single calf is born after a gestation period of a little more than nine months. The mother leaves the herd for a few days to give birth, and the period alone with the calf serves to cement the bond between the two. The calf is weaned after a further nine months.

Standing up to tigers

Threats to the gaur population include habitat loss, hunting and diseases transmitted by domestic cattle. Young gaur are also vulnerable to attacks from tigers. Usually gaur will take flight from such formidable opponents, but occasionally they stand their ground. With their considerable bulk and impressive horns, they can be imposing animals, and a herd will sometimes face up to a tiger by forming a defensive ring. Despite this tactic, tigers frequently kill calves and will also attack solitary adults.

At the first sign of danger, a gaur raises its head and sniffs the air; it has a keen sense of smell and is able to detect a tiger at a range of 1300 ft. If the danger is close at hand and the gaur needs to escape, it will utter a sharp sound like a trumpeting sneeze, then leap away, pounding its forefeet down together on the ground as an alarm signal.

A subspecies of the gaur, known as the gayal, or mithan, has been domesticated. Many of these domestic animals have returned to the wild to form feral populations. The gayal is smaller than the gaur, and a large male rarely weighs more than 1500 lbs. Its shape is more thick-set and compact; its limbs are shorter, and the hump on its back is not so pronounced. As with many domesticated animals, its coat varies considerably in color, and albino and dappled varieties do occur. The gayal is normally kept as a source of meat and is rarely milked or used as a draft animal. Many hybrids have been created from crosses with both domestic cattle and wild gaur.

The cow-like banteng

The banteng of Southeast Asia and the Indonesian islands is smaller and lighter than the gaur, growing to 5 ft. 6 in. at the shoulder and weighing 2000 lbs. It is a more slender animal, and the hump of the male is less pronounced. In shape it resembles the domestic European cow. Its coat color ranges from red-brown

ABOVE The white cattle of Chillingham Park in northern England are descendants of domestic oxen. The cattle have roamed free as wild animals in the land surrounding Chillingham Castle for over 700 years.

BELOW The hierarchy of a herd of cattle can often be seen from the position of each animal. In this herd of Highland cattle, the dominant animals graze in the center while their subordinates stand guard around the perimeter.

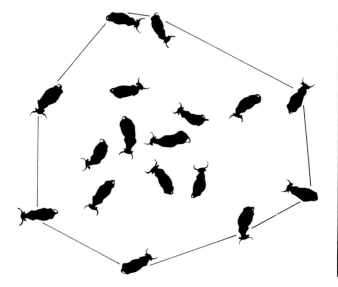

ABOVE **Domesticated Camargue cattle roam freely over the expansive wetlands of the Rhone** delta in southern France, living in much the same way as their wild ancestors before them.

wild cattle are driven further and further back into the forest, their range continually shrinking. Like the gaur, they are also vulnerable to the diseases carried by domestic cattle.

The banteng occupy similar habitats to those of the gaur, preferring thick forest with open clearings, where they can graze the grass and browse on low-growing foliage. These conditions often exist in upland forests, and banteng will climb to an altitude of 6500 ft. above sea level to find suitable feeding areas. In common with many other wild cattle, they are most active at night. During the day they tend to stand still, chewing the cud in the shade beneath trees or among dense bamboo. In the evening, as darkness falls, they go out into the clearings to forage for leaves and grasses. They may graze all night, slipping back into cover at dawn.

The docile banteng has been domesticated in every part of its range and has been introduced as a domestic animal in northern Australia, where it has escaped to form feral populations. Domestic banteng have been crossed with other cattle, especially the Asian humped breed known as the zebu. Wild banteng also mate with domestic animals, so that unofficial hybrids are common. These hybrids may be kept as domestic animals or they may return to the wild, providing a continual interchange of blood. Although animals like the banteng may not die out in the wild, it is possible that they will gradually change through repeated hybridizations. One of the advantages of a nature reserve is that it blocks this process and keeps the wild strain pure.

Endangered kouprey

Of all the wild cattle, the most threatened is the kouprey of Southeast Asia. It is one of the world's rarest mammals and is on the brink of extinction; only occasional sightings of the animal are made. The kouprey is halfway between a gaur and a banteng in size, with a grayish coat and pale gray "stockings." Females have distinctive lyre-shaped horns, while mature males have widespread antlers with frayed tips. Like the gaur and the banteng, the kouprey lives in dense forest with open glades—and until 1937 it kept itself so well hidden that it was quite unknown to Western zoologists. Up to this time, the type of shifting agriculture practiced by the local forest farmers may have benefited the animals, since it created abandoned clearings that were ideal for feeding kouprey.

in females and immature males, to a rich dark chestnut in old bulls. It has white "stockings" on its legs, a white rump patch and a white tip to the muzzle. Its horns are fairly short and upswept and, in bull banteng, the area between the horns is hairless.

The four races of the banteng occur as isolated populations. The Javanese banteng is now rare: it survives in the wild in Java, in certain protected areas within the Udjung Kulon and Baluran reserves. The banteng of Borneo is also endangered and is found in reasonable numbers only in the Kinabalu National Park, while the race belonging to the island of Bali may now have been totally domesticated as the Balinese ox. The continental subspecies, known as the Burmese banteng, or tsaine, once ranged throughout Southeast Asia but now survives only as isolated populations in patches of undisturbed forest such as the Pidaung Reserve in Burma and the Khao Yai National Park in Thailand.

Where banteng are in direct competition for grazing land with domestic cattle, herdsmen will ensure that their domestic stock take precedence. Inevitably, the

RIGHT **Protected from the biting Himalayan wind by a thick coat of matted hair and kept warm by the fermenting grass in its stomach, the yak manages to survive in one of the most unforgiving** habitats in Asia—the cold deserts and alpine tundra of the high Tibetan plateau.
BELOW RIGHT **Two young male yaks spar with each other to establish their place in the herd.**

A victim of war

Had the actions of man continued to prove beneficial, the kouprey population might have remained stable. Unfortunately they did not, for the animals live in the forested lands of northeast Kampuchea on the border between Laos and Vietnam—an area that was devastated during the Vietnam War. Much of their habitat was destroyed, and subsequent conflicts in the region have increased the kouprey's plight. In 1969, the population estimate stood at 100—since then there have been only occasional sightings of the animal.

The few kouprey that still survive in the wild live in small herds numbering up to 20 animals. More than one mature male may be present in each herd. The mating season falls between April and May, and a single calf is born after a nine-month gestation.

Forerunner of domestic cattle

The many varieties of domestic cattle differ greatly in appearance, but they can all be interbred to produce fertile offspring. All domestic cattle are regarded as one species and descendants of the same animal: the original wild ox, urus or aurochs.

Now extinct, the aurochs was well known in ancient times and is mentioned in the work of the Roman author Pliny. It once lived throughout Europe and Asia, though it disappeared from Asia about 2000 years ago. However, it remained common in the forests of western and central Europe, at least until the late years of the Roman Empire and probably well into medieval times. Thereafter its numbers declined owing to hunting and the destruction of its habitat as land was cleared for forestry and agriculture. The animal became extinct in the British Isles by the 10th century AD, and from the 15th century onward it became increasingly rare in continental Europe. It finally took refuge in remote forestland in Poland. The last individual died in 1627, in the Jaktorow forest in Masowia, near Warsaw.

Although the aurochs itself has vanished, numerous descriptions and illustrations remain, and these inspired a team from Munich Zoo to attempt a reconstruction of the primitive ox. The team selected domestic breeds that still displayed primitive characteristics, crossbred them, and inbred the offspring to produce animals resembling the wild ancestors. Though the experiment was successful in recreating the build and coat color of the original aurochs, the inbred animals were much smaller than their real ancestors. Examples of these reconstructed aurochs can be seen today in zoos and wildlife parks.

The true aurochs were large-horned animals that could reach a ton in weight. There was a marked difference between the two sexes. The males were not

ABOVE Its curved horns gleaming in the sun, a big bull yak chews the cud. Formerly widespread over the foothills of the Himalayas, the wild yak has been driven out of the lush grasslands and up into more barren, inaccessible areas where the climate is harsh and the grazing poor. As a result, the birth and survival rate has suffered, and the yak is now on the list of endangered species.

only larger but also darker in color, with yellowish stripes along their backbones (this pale back stripe persists in many domestic breeds of cattle, such as the Longhorn and the Hereford). Unfortunately, zoologists know little about the aurochs' biology and behavior; evidence is scarce, and the evidence that does exist from classical and medieval authors lacks scientific credibility.

Although the aurochs appears to have been the ancestor of all domestic cattle, the variety of breeds in existence around the world suggests that domestication took place in more than one area, using local varieties of the wild oxen. The qualities that local farmers considered valuable in the cattle also varied, and a number of distinct types of domestic cattle were developed in different places. The Spanish fighting bull and the black Camargue ox of southern France are considered to be among the most primitive breeds and are probably very close to the aurochs in their structure and habits. Other breeds reminiscent of the aurochs in habit, shape or color and patterning of their coats, are Corsican oxen, the cattle of the Hungarian and Ukrainian steppes, Scottish Highland cattle and the English White Park cattle. Some of these breeds were used in reconstructing the aurochs.

Humpbacked cattle

The most distinctive race of domestic cattle is the humpbacked variety of Asia and Africa, known as the zebu, or brahman. The race includes the sacred cattle of India. Zebu differ in many ways from typical European cattle. They are adapted to hot climates, possessing well-developed sweat glands and loose skin that increases their surface area and enables them to lose heat more easily. Their skulls are unlike those of European breeds, and the makeup of their blood is also different. All of this might suggest that zebu are descended from a different species—one of the wild humpbacked Asiatic species such as the gaur or the banteng. However, analysis of their bone structure indicates otherwise. Their skeletons are far more like those of European cattle, and whereas the hump of an animal like the gaur covers bony extensions of the spine, the zebu's hump consists of muscle and fat. The zebu is probably the descendant of a local subspecies of the aurochs common in India in ancient times.

Zebu originated in India and were taken to Asia Minor and then to Africa, where they gave rise to a number of local breeds. Notable among these is the watussi that grazes on the savannahs of East Africa. The cows have horns up to 5 ft. long. The zebu has also been crossed with other cattle in an effort to combine its tolerance of hot climates (and resistance to tropical diseases) with the high growth rate of European breeds. The most successful of these zebu crosses, the Santa Gertrudis, was developed in Texas to suit the hot, dry conditions of the southern USA. Zebu bulls were crossed with stock originating from Herefords, Texas Longhorns and Shorthorns. The Santa Gertrudis was first recognized as a breed in 1940 and has since spread to Cuba, South America and Australia.

Although breeds such as the zebu and its relatives are able to cope with high temperatures, most domestic cattle are better adapted to cold climates. As ruminants, they digest grass by storing it initially in a large stomach chamber called the rumen. Here it ferments, and in the process generates a lot of heat. Regardless of the outside temperature, the contents of the rumen simmer away at a steady 104°F, providing the animal with its own internal heat source. Some members of the wild cattle family are able to live in areas with a harsh climate because they have this internal source of warmth. Of these, the most remarkable is the yak.

On the roof of the world

The yak lives on the icy wastes of the Tibetan plateau at altitudes of up to 20,000 ft. above sea level. It is well equipped for such a hard life. The animal's body heat is conserved by a thick, woolly undercoat. On top of this it has a protective outer layer of long, coarse, dark brown hair that hangs down about its legs and over its forehead in shaggy, matted fringes. It is a massively built animal—a bull may weigh up to 2200 lbs. and measure over 6 feet high at the shoulder. The yak has a large hump like the gaur, a drooping head and long, curved horns.

The yak is not an energetic animal—inaction is, in fact, part of its survival strategy. By cutting its energy expenditure down to a minimum, it makes the best use of the scarce, often poor-quality mountain grazing. Kept warm by its fermentation system and insulated by its massive bulk, thick hide and dense coat, the yak can survive the most terrible snowstorms out on the open mountains, turning its back to the blizzard and placidly chewing the cud as its world turns white.

Despite their protection from the cold, yaks usually spend only the warmer months (August to September) at altitudes where there is permanent snow cover. For the rest of the year, they migrate to lower heights to feed on grass, herbs and lichens.

The yak has been domesticated for centuries, and in the past it was a mainstay of the Tibetan way of life. The animals provided transportation, power, meat, leather, wool, milk and even fuel. In a part of the world where firewood was hard to come by, the dried dung of the yak made an excellent substitute. Domestic breeds of yak occur throughout Tibet, spreading south to Nepal and Bhutan and northeast as far as Inner Mongolia. At altitudes over 6500 ft., the yak is the only pack animal (it can carry up to 350 lbs.) and is the only mount available. Horses and oxen cannot function effectively in the thin air and harsh climate.

ABOVE Heads down, a pair of plains bison crop the grass and brushwood on the American prairie. These bison are essentially grazing animals and are well adapted to life on the open grasslands. At one time they roamed the prairies in vast herds. The animals were followed by nomadic tribes of American Indians who were dependent on them as their main source of food, clothing, footwear, tents, bowstrings and many other items of daily use.

The range and population of the wild yak have diminished so much that the wild yak is now considered an endangered species. As with other wild cattle, it is hard to establish whether the yak's worst enemy is the hunter with a rifle, or the rancher who breeds domestic cattle that compete for grazing. Certainly the wild yak has been forced out of the best pastureland into marginal areas, where its food supply is poorer and its ability to reproduce successfully is reduced. Nineteenth-century authors have described how the mountain pastures were once black with the grazing herds. The Russian naturalist-explorer Przewalski saw herds of "several hundred and even several thousand head"—today such a gathering would represent a large portion of the world population. These herds would have been composed of females and young; the males tend to live alone or in small groups.

THE EUROPEAN BISON
— SAVED FOR POSTERITY —

The story of the European bison or wisent shows how determined action by naturalists and animal breeders can save a species that seems doomed to extinction.

Bison were once found in forests across Europe and Asia, but hunting and the clearance of the forests for timber and agriculture led to a steady reduction in their numbers. In Russia, the dwindling bison population came under the protection of the czars from 1803, but World War I and the Russian Revolution brought this to an end. In 1865 the Duke of Hochberg established a breeding center in the Pszczyna forest in Upper Silesia, a part of southern Poland then under Prussian rule. The herd flourished and numbered some 70 animals by 1921—but when law and order broke down as a result of political upheavals later that year, the animals were slaughtered wholesale, and only three survived.

Meanwhile, the last surviving truly wild herd, living in the primeval Bialowieza forest on the border between Poland and the USSR, had met a similar fate. In 1914 the herd numbered 727; by 1919 they were all dead, killed for meat during World War I.

Poles to the rescue

In a last-ditch attempt to save the species, the Polish zoologist Jan Sztolcman founded The International Association for the Preservation of Wisent in August 1923. The Association commissioned a report on the current position of European bison. The findings did not make encouraging reading. The total population of the animals had been reduced to a mere 66 animals, scattered among the zoos and parks of Europe: they comprised 24 adult males and 22 adult females (three of each being unfit for breeding) plus 20 calves.

In 1929 the three surviving animals from the Pszczyna herd were established in a fenced-off part of the Bialowieza forest, where they were joined by three more bison from zoos in Sweden and Germany. By 1939 their numbers had swollen to 30. The herd survived World War II intact, and by 1946

478

there were 46 individuals. Elsewhere, the species had been hit much harder by the war (only 12 survived in Germany), and in 1952 the Association decided to establish a self-supporting herd in the forest. The experiment proved a success; the herd flourished and by 1983 numbered over 250. Meanwhile, several more breeding centers had been set up in the USSR and Romania, with similar success.

A primeval refuge

Conditions in the Bialowieza primeval forest—in part, a major European nature reserve—are ideal for the European bison: pine or mixed forest, with large clearings and plenty of undergrowth for the bison to browse on. Like the wood bison of America, the European bison is a browser rather than a grazer and eats mainly leaves, twigs and bark.

ABOVE Slimmer and more agile than its American counterpart, the European bison or wisent is a native of the open, mixed primeval forests of northern, central and southern Europe. Driven out by hunting and agriculture, wild European bison now live only in the Bialowieza forest in eastern Poland and the Caucasus National Park, USSR.

FAR LEFT A family group of European bison grazes the open hillside in the Caucasus National Park.

LEFT Although it has relatively poor sight, the bison is a wary, alert animal, quick to pick up the scent of an intruder.

BELOW A bull marks a tree at the limits of its territory.

479

over six feet at the shoulders. The forequarters are heavily built, it has a pronounced hump (supported by extensions from the spine), a heavy, low-slung head and a shaggy coat. The male, in particular, has long, thick hair on its forelimbs, and a dense mane on its head, neck and shoulders. By contrast, the hair on the hindquarters is short and sleek, giving the animal a top-heavy appearance.

The most impressive animals are those that formerly lived in vast herds on the North American prairies—these are now regarded as a separate subspecies known as the plains bison. The other subspecies, the wood bison, is slightly larger and darker, with more balanced proportions, wider hooves and longer horns. It is closer to the European species in form. Its habitat—open forest—also resembles that of the European bison, which may explain why the two have remained so alike. Indeed, they are so similar that they can successfully interbreed.

At the end of the 19th century there were perhaps as many as 60 million American bison, widely distributed over the prairies and woods between the Rocky Mountains and the East Coast, north to the Great Slave Lake and south to Mexico. The vast herds—often numbering several hundred thousand—were hunted by the Plains Indians, who followed them as they wandered in search of pasture. Indian culture depended on the bison, although Indians never attempted to domesticate them in any way; there was no need.

Indians killed an insignificant number of the bison population—their weapons were primitive and their requirements were modest. The situation had changed by 1800, with the arrival of European settlers and their firearms. Almost immediately the pioneers started to kill off the bison. At first the killing was done at random, to obtain meat and defend the grazing land set aside for domesticated cattle. But in time it developed into a systematic massacre, intended partly as a means of depriving the Indians of their chief economic resource.

The mating season normally falls between September and October. Male yaks fight fiercely for domination of the herds, and may inflict numerous wounds. They utter strange grunting calls, which can be heard from domestic yaks all year round. The gestation lasts for about eight-and-a-half months, and the young yak is protected and guided by its mother for at least a year.

The imposing bison

The American bison and the European bison, or wisent, are similar in appearance. It is likely that bison from the European stock migrated into America across the land bridge that once joined Siberia to Alaska. The migrants were cut off when the Bering Strait opened up at the end of the Ice Age. Once isolated from each other, the two populations developed differently to suit their respective environments. Today they are sufficiently distinct to be classified by most zoologists as separate species, although some still regard the American and European bison merely as subspecies of the same animal.

The American bison is the only truly indigenous member of the wild cattle family in the New World. It is often referred to as the "buffalo," although it is not a true buffalo like those of Africa and Asia. It is an imposing beast, weighing up to a ton and standing

STALWART GRAZERS

Killed for fun

The availability of cheap, efficient firearms and the construction of a railway between the Atlantic and the Pacific coasts sealed the fate of the bison. From 1865 they were being shot from the windows of trains, just for fun. Between 1871 and 1875 all the bison living south of the transcontinental railroad were exterminated by hired hunters, and within 20 years it seemed that the herds living further north were likely to follow. The American bison was on the verge of extinction. By New Year's Day, 1889, there were only some 300 wild bison in the United States, 550 in Canada and some 400 kept in zoos or on farms—and 40 of these were hybrids.

Fortunately, in 1905, the society called "Friends of the Bison" was launched, and the outlook for the species began to improve—though not before an epidemic had threatened to wipe out the few remaining wood bison. Once this epidemic had been dealt with, the society was able to encourage the conservation and breeding of the surviving herds, and today the threat of extinction has receded.

The bison population is dispersed among several parks, reserves and even private farms, where the animals live more or less wild (although they are supplied with fodder in winter). Two of the wildest refuges for the bison are Yellowstone National Park in the USA and the Wood Buffalo National Park in Canada, where a small herd of pure wood bison—only discovered in 1957—survives among 12,000 hybrids. There, the bison still live as they did in the past. On the whole, the survival of the American bison seems to be assured, and the total population is now estimated to be 50,000. In some local areas where the habitat is restricted, there is even an overpopulation problem and controlled hunting has to be carried out.

Social life

Mature male bison live alone or gather in small, unstable bands at the edge of herds. They only move in among the females during the mating season (July to August). Females, on the other hand, are never found alone: together with young males (up to three years old) they form groups containing at least five members, usually more. The groups join together to form larger herds during the mating season. Herd size depends on the density of the local population: in some areas the average herd may contain as many as 100 animals, while in others there may be only 20.

The dominant bulls have to fend off many rivals during the mating season, and skirmishes between males often develop into violent fights. A threatening bison may stand sideways to his opponent, with arched back and head slightly lowered. In this position, he may bellow several times. If neither animal will give way, they may begin to fight. Fighting techniques include charging each other head-on, often at considerable speed, and butting one another in the flanks. When the bison clash head-on, the force of the impact depends on speed and weight, but fighting flank-to-flank is a matter of strength—the two males use the massive muscles of their necks and forequarters to power each blow.

The padding of thick hair on the bison's forehead softens the impact of clashes, while the mane provides protection on the flanks and neck. Nevertheless, the battles for supremacy often result in injury, and deaths are not unknown. To minimize the damage, bison employ a degree of ritualization when they are fighting: it is not uncommon for an opponent who finds himself in a position of advantage, where he could easily strike his rival in the flank, to stop short and wait for the other to resume his fighting stance.

A male bison has other threat displays, in addition to the sideways stance, which may be sufficient to dissuade a rival from fighting. An aggressive bull bison will bellow loudly, stamp at the ground and snort, producing a sound similar to a sneeze. This is also a signal of alarm and suspicion. The most spectacular display is ritual wallowing in the dust—normally done

ABOVE A bull bison paws at the ground before charging at his rival. These head-butting fights are awesome; during the mating season, 50 or 60 such battles may break out in the herd at the same moment.

481

to eliminate parasites. A bull wishing to make a great show will throw itself on the ground and roll over from side to side, covering its coat in dust and whipping up a large dust cloud. The animals often return to the same wallowing areas, wearing a bare oval patch on the ground up to 65 ft. in length.

ABOVE A small herd of bison grazes at the edge of a conifer forest in the American Northwest. Before the bison were nearly wiped out by hunters and settlers during the 19th century, a big herd could number several hundred thousand animals. Today a large herd rarely contains more than a hundred animals, and most are much smaller.

Back from the brink

The European bison, or wisent, is the largest land animal in Europe. It is very similar in appearance to the American bison, but has longer hind limbs, thicker, shorter and blunter horns, and shorter hair on the neck and shoulders. Its overall size is similar—the animals weigh up to 2000 lbs., with a slightly smaller shoulder height of up to 6 ft. European bison are woodland animals, preferring open forests with plenty of undergrowth where they can browse on deciduous trees and shrubs in summer, and evergreen trees and shrubs in winter. By 1919 the species had been wiped out in the wild, and escaped extinction only because a few specimens (66 in total) survived in zoos and parks.

All the surviving animals belong to the Lithuanian (lowland) subspecies; the Caucasian (highland) subspecies became extinct in 1925. Its main features have been "reconstructed," rather in the manner of the aurochs, and the resulting animals introduced to the Caucasus National Park in the USSR.

ANTELOPES OF SANDS AND SAVANNAH

Among the grazing antelopes, the gnus migrate
in the thousands across the African plains,
while the desert oryxes sport nature's most
spectacular horns

White-tailed gnu

Sable antelope

Bontebok

Addax

Lichtenstein's hartebeest

Arabian oryx

Within the bovid family, some 23 species are classed as grazing antelopes. They are divided into three tribes—the hartebeests, gnus and their close relatives; the horse-like antelopes (which include the oryxes and the sable antelope); and the reedbucks, waterbuck and their relatives (described on pages 503-516).

Grazing antelopes occur throughout Africa, except the far north, wherever grass grows thickly enough to provide an adequate food source. They have colonized most types of habitat, from flooded swamps on the margins of lakes and rivers to the arid desert soils of the Kalahari and the Sahara. Although normally barren, the deserts can sprout a lush growth of grasses after a rainstorm. The antelopes grouped with the gnus and the hartebeests are generally large animals with long heads, high shoulders and sloping hindquarters; both sexes have horns. They have scent glands beneath their eyes (preorbital glands) and between the toes of their forefeet (pedal glands). Males and females have similar markings, and in some cases they are almost identical in appearance.

The swift hartebeests

Hartebeests have a wide distribution over most of Africa, south of the Sahara. The two species—the hartebeest and the less widespread Lichtenstein's hartebeest—have long, narrow heads, high bony bases for their ridged horns, and sharply sloping backs. They may weigh up to 500 lbs., and stand 5 ft. in height at the shoulders. They are fleet of foot, strong and vital, and it was these qualities that gave the animals their name—"hartebeest" is a Boer word that may be translated as "tough beast." The Boers found that it was almost impossible to hunt the animals on horseback in the way they hunted other antelopes, and a single shot was rarely sufficient to bring them down. Nevertheless, the persistence of hunters eventually made its mark, and hartebeest populations have declined over much of Africa.

The hartebeest has 11 distinct subspecies ranging over the grasslands of Africa. Until recently there were 12: the northern hartebeest or bubal hartebeest is now extinct, but small groups of these animals survived north of the Sahara until about the 1930s. Several of the subspecies are anatomically closely related to the extinct animal. They range across the continent in a belt immediately south of the Sahara, from Senegal to

TOP A female hartebeest keeps watch over her calf on the African savannah. The calf stays hidden in long grass until it is two weeks old; fed and kept clean by its mother, it rarely moves and gives off very little scent so as not to attract predators.

ABOVE Able to spring suddenly high into the air, hartebeests are agile, swift-footed animals.
PAGE 483 The great trek: part of a large herd of brindled gnu, several hundred strong, migrating to more productive feeding grounds.

485

Chad. Among these is the western hartebeest, a large animal with a sandy brown coat and dark markings on the front of its forelegs. It has massive horns that form a U-shape at the base when viewed head-on.

The Lelwel hartebeest, the Chad hartebeest, Roosevelt's hartebeest and Jackson's hartebeest belong to a second group of subspecies found in central Africa and parts of Uganda and Kenya. They can be recognized by their elongated heads, uniformly reddish-brown coats and the V-shape made at the base of their horns. Swayne's hartebeest, native to eastern Ethiopia and Somalia, is a chocolate color, but adults take on a silvery hue as the tips of the hairs whiten with age. The tora hartebeest, which survives in a few areas in north and west Ethiopia and the eastern Sudan, has a light brown coat. Like Swayne's hartebeest, it is a fairly small subspecies with horns that curl forward and outward before coming closer together at the tips.

Precipitous decline

When first recorded in 1891-1892, Swayne's hartebeest was abundant, with individual herds of over a thousand strong recorded on the plains of Somalia. But within a few years the vast herds were no more, and by 1905 scarcely 880 survived. The precipitous decline was the result of two factors—disease and hunting. The epidemics of cattle disease that swept through much of Africa at the end of the 19th century had a particularly severe effect on the hartebeest populations. However, they would have recovered fairly rapidly were it not for uncontrolled hunting carried out by the European soldiers fighting in Somalia during the early 1900s.

To make matters worse, the Ethiopians took advantage of the administrative chaos reigning in Somalia and sent hired poachers into the country on raiding expeditions. The poachers were so successful that they reduced Swayne's hartebeest to the verge of extinction in Somalia, and since then the subspecies has had little chance to reestablish itself.

The red (or Cape) hartebeest lives in the dry regions of southern Africa. It is particularly handsome, with a chestnut-red body and black and white markings on its head, belly, rump and legs. At one time, enormous herds of red hartebeest gathered on the plains when seasonal rains produced an abundance of pasture, but such scenes will probably never be seen again. Their numbers declined rapidly in the second half of the 19th century as the land was settled by European farmers anxious to preserve the grazing for their own livestock. They became extinct in the wild in South Africa and

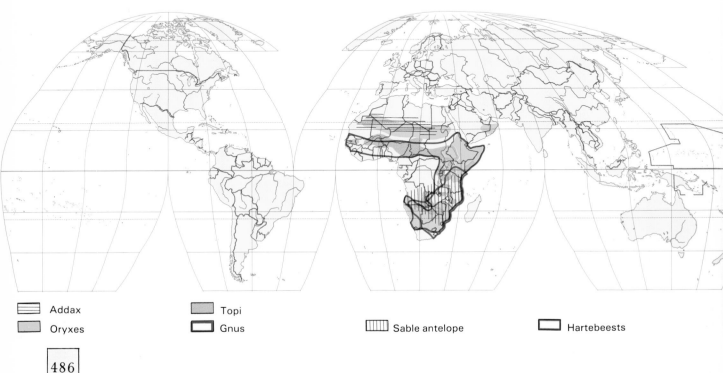

Addax

Oryxes

Topi

Gnus

Sable antelope

Hartebeests

disappeared from many other areas where they had been numerous. Fortunately, measures to protect them were taken in good time, and today the danger of extinction seems to have been averted. They are still numerous in Botswana, and they have been reintroduced into parks and nature reserves in South Africa and Zimbabwe.

Coke's hartebeest

Coke's hartebeest, otherwise known as the kongoni, is still common in much of southern and central Kenya and northern Tanzania, although it is declining in some places, especially on the edge of its range. A sandy-colored animal with short, widely spread horns, it lives on dry, grassy or partly wooded plains at altitudes up to 6500 ft. It feeds almost exclusively on grass, and can survive on very little water compared to most other ruminants—an adaptation it has made to its hot, dry environment.

The female Coke's hartebeest normally lives in small groups that range over areas of two square miles or more, while individual males defend small territories of, on average, one-tenth of a square mile. The much larger female home ranges may encompass the territories of many males. Rival males are always chased out of territories, but a female is welcome in

ABOVE Two females and one young Coke's hartebeest, also referred to by their Swahili name ''kongoni,'' in their typical dry grassland habitat. Coke's hartebeest feed mainly on grass, and as an adaptation to their arid environment, they are able to survive on a limited supply of water.
FAR LEFT The map shows the distribution of several species of the grazing antelopes in Africa and the Arabian Peninsula.

any area defended by an adult male. He will make every effort to encourage her to stay on his patch for as long as possible.

The young are born in March or September after a gestation period of eight months. In each case the birth occurs at the beginning of a rainy season, so the young are ideally placed to benefit from the fresh young grass that appears when the rains come. New grass is rich in vegetable protein, boosting the mother's milk yield as well as providing the young with a source of high-quality solid food just when it is most needed.

At calving time the pregnant female leaves the herd and seeks the cover of scrubland. She may take her most recent offspring with her. In this retreat, concealed from the eyes of potential predators, she gives birth to a single calf. In less than half an hour the calf is able to stand, and a few minutes later it is being

ABOVE An adult (left) and an immature western korrigum still with its short horns. The western korrigum is a subspecies of the topi, found in West Africa. It normally lives in small family groups, but during periods of drought these groups may gather into large herds to feed on remaining pasture.

suckled, but it does not follow its mother as she grazes. During the first two weeks of life the calf lies hidden in the long grass and only has contact with its parent at feeding times.

The young Coke's hartebeest is at great risk during this period: alone and defenseless, it is vulnerable to predators that pass by. To minimize the danger of discovery, the scent glands of the young animal remain inactive while it is lying in the grass, and its mother licks away any urine and feces it has excreted. During the first two weeks of life the calf, therefore, gives off very little scent. The mother also eats the placenta immediately after the birth, eliminating another source of scent which might guide a carnivore to the spot. The calf lies almost completely immobile, avoiding any rustling or obvious movement that might give it away. If a predator approaches in spite of these precautions, the young hartebeest may be agile enough to run away and make its escape, even when it is only a few days old.

Lichtenstein's hartebeest

Formerly classed with the other hartebeests as another subspecies, Lichtenstein's hartebeest is now considered to be a quite distinct species. It is distinguished from the other hartebeests by a bright reddish back, which contrasts with the creamy-fawn color of the rest of its coat. Its horns are short, flattened and thick at the base.

Lichtenstein's hartebeest lives mainly in the wooded and scrubland areas of eastern, southwestern and southern Africa. Not so gregarious as some of the other hartebeests, it lives in small herds of up to 10 animals. Each herd is a "family" group consisting of a dominant adult male, a number of females and their young. Unmated males, which are normally younger animals, form separate bachelor herds, while elderly males often lead solitary lives.

The dominant males are aggressively territorial in their habits. Any adult male that has a harem will strenuously defend the area carved out as a territory and will not tolerate the presence of rival males. He does not always lead the herd, however; if they are attacked, the whole herd will follow the oldest female as she takes flight. The adults have acute senses of sight and hearing and, although usually silent, they may snort through their nostrils when alarmed. The young are born in July and August. They are vulnerable to attack from predators, and many fall victim to lions, hunting dogs, cheetahs and leopards.

The topi

Topis are agile, robust animals with horns that may grow to lengths of 24 in. or more and a body weight of between 285 and 375 lbs. Height at the shoulders varies from 44 to 50 in. Topis are found over most of Africa south of the Sahara. There are seven subspecies: the topi itself, the tiang, the jimela, the tsessebe and three subspecies known as korrigums.

The tsessebe (or sassaby) inhabits wet, thinly wooded grasslands from Zambia south to northern Botswana, the Transvaal and northern Natal. It is a large antelope about the size of a red deer, with a glossy chestnut-red coat, white belly, blackish markings on its face and forelegs, and lyre-shaped horns. It lives in small family groups, but sometimes these groups join together to form large herds of up to 200 individuals. Large herds usually form during periods of drought, when the animals congregate in areas where grass (their main food) is still available. The tsessebe is among the fastest of all African antelopes and can run at speeds exceeding 45 miles per hour. One tsessebe may leap over another's back to escape if suddenly alarmed.

The topi itself is also a fast runner. Slightly smaller and a lot lighter than the tsessebe, the topi lives mainly on the savannahs of East Africa, but there are a number of distinct races which differ from one another in their choice of habitat. Although its distribution is not extensive, the topi is plentiful in many parts of its range, and herds of several thousand animals have been recorded. Generally these large herds gather in the dry season (usually from June to October), when the animals migrate to new pastures; in normal circumstances the topi lives in groups of 30-40 animals, each group containing females and their young, and usually one mature male. Topis commonly graze in the company of other antelopes, particularly brindled gnus, and may also be seen with plains zebras and ostriches.

Horns to the ground

Both male and female adult topis often behave in a curious way, vigorously rubbing their horns against the turf, kneeling on their forelegs and even throwing large clods of earth in the air. The most likely explanation for this activity is that it displays the animal's physical energy, helping to establish its rank within the herd.

The reproductive cycle begins in December with the rutting season. Up to 100 males may congregate together on traditional breeding sites. Each male attempts to defend a small "stamping ground," the size of which depends on the density of males at the site. Stamping grounds may be as small as 80 ft. in diameter if there are a large number of males in a relatively confined area.

A male moves among females in his territory with head held high and muzzle stretched out in front. He will chase away any other males of lower rank, and will persistently try to keep the females from straying. If a rival refuses to run away when threatened by the resident male, the two may fight. They begin by circling around each other, issuing threats of increasing intensity, and end up battering each other with their horns. Eventually the weaker animal is forced to submit, give ground and leave.

ABOVE Mature male blesbok (a subspecies of the bontebok) have distinctly darker coats than the females. Males are aggressively territorial if they have a harem to defend, and will chase off any male intruders who dare to cross the scent-marked territorial boundaries.

The young are born seven or eight months after mating, often at the start of the rainy season, and this allows them to take advantage of the new grass in the same way as young Coke's hartebeest.

South African survivors

The bontebok belongs to the same genus as the topi but is smaller and more lightly built, weighing 130-220 lbs. and standing 35-45 in. at the shoulders. It has large, distinctive patches of white on the face, belly and legs. Adult males are bigger and darker than females and, apart from the white patches, the coat is a dark chestnut-brown. The neck and hindquarters are darker, with a purplish sheen. Bontebok live only in the open grasslands of southern Africa.

There are two subspecies: the bontebok itself and the blesbok. The bontebok is bigger, darker and has a large white rump patch, whereas the blesbok has a dark marking on its face, breaking the white facial patch, and a small pale yellowish-brown area around the root of the tail. They both have rather small horns that are similar in shape to those of the topi with 10-20 widely spaced ridges.

The blesbok is widespread in game reserves and on ranches across South Africa, although in the 19th century it was almost wiped out across its range by indiscriminate hunting. Its numbers have increased since a number of private landowners banned hunting of the animal on their land. The bontebok is one of the rarest ruminants in southern Africa and has been on the verge of extinction several times. By 1931 indiscriminate hunting had reduced its numbers to a mere 30 or 40. Just as the species seemed likely to vanish off the face of the earth, the first Bontebok National Park was established near Bredasdorp in the Cape Province, South Africa, and the surviving animals were introduced there. Protected from the hunters, they bred and flourished, and later, in 1960, a new Bontebok National Park was set up replacing the first one. This was sited near the town of Swellendam, 45 mi. north of Bredasdorp, in a richer pastoral area, a habitat more suited to the bontebok. The initial tiny population increased to almost a thousand in just 10

years. As a result of this policy the bontebok is no longer such an endangered species, although its distribution is still very limited.

Bontebok usually live in small groups. Some of these are made up entirely of young males; others consist only of females and their young. Active breeding groups include a number of females, their young and one adult male. Bontebok have a rigidly timed breeding season, unlike most other African ungulates, and they always mate between January and March; the calves are born in September and October. Unlike the hartebeests, the female does not withdraw from the herd to have her calf, and she does not leave the calf to fend for itself in the long grass. Active from birth, the calf is immediately able to follow its mother around as she grazes, and so enjoys the protection of the rest of the herd. Even so, the calf spends most of its time resting during the first two weeks of life, and is suckled by its mother every hour or two.

A group of females with young range over an area of between 10 and 70 acres. There are rarely more than nine adults to each group, and the size and composition of the party tend to remain constant for

GRAZING ANTELOPES
CLASSIFICATION: 1

The grazing antelopes belong to the subfamily Hippotraginae—a part of the large and diverse family Bovidae. With the sole exception of the Arabian oryx, they are all native to Africa. The subfamily is divided into three tribes: the Alcelaphini, which includes the gnus and the hartebeests; the Hippotragini, also known as the horse-like antelopes, which includes the oryxes and the sable antelope; and the Reduncini, which includes the reedbucks and the waterbuck.

Alcelaphini

Members of the tribe Alcelaphini are medium-sized to large antelopes, with high shoulders and long, narrow heads. Both males and females possess horns. There are eight species, some of which are broken down into several subspecies.

The genus *Alcelaphus* contains the hartebeest, *A. buselaphus*, which ranges throughout sub-Saharan

Africa outside the forested zone, and Lichtenstein's hartebeest, *A. lichtensteini*, of eastern, southern and southwest Africa. The hartebeest has 11 subspecies, including the Coke's hartebeest or kongoni, of Kenya and Tanzania.

Two species belong to the genus *Damaliscus*—the topi, *D. lunatus*, which has a similar range to that of the hartebeest, and the bontebok, *D. dorcas* of southern Africa. There are seven subspecies of the topi, while the bontebok has two subspecies—the bontebok itself and the blesbok. Only one species makes up the genus *Beatragus*—the hirola or Hunter's hartebeest, *B. hunteri*, of eastern Kenya and southern Somalia.

The gnus or wildebeests belong to the genus *Connochaetes*. The white-tailed gnu (or black wildebeest), *C. gnou*, is confined to South Africa, but the five subspecies of the brindled gnu (or blue wildebeest), *C. taurinus*, range over grassland and open woodland throughout eastern and southern Africa.

ABOVE **Born within the confines of a zoo enclosure, this young blesbok may yet experience the freedom—and danger—of the open savannah. Captive breeding programs often play a key role in the preservation of animals such as the grazing** antelopes that are **threatened throughout their range by hunting, habitat destruction and competition with domestic livestock.**
RIGHT **A male blesbok raises his foreleg to a female, his neck and tail outstretched in a courtship gesture.**

many months. The male groups are much larger, with up to a hundred or more individuals. The age range varies, but most of the animals are immature or sub-adult, and there are few mature males. These are open societies—an animal may withdraw from the group at any time, or the group may break up into smaller units. Males may even be joined by immature females.

Nomadic bands of bontebok

The male groups do not have a fixed home range like the female groups; instead, the males lead a nomadic life, wandering over the landscape with no need to take heed of territorial boundaries. Since they have no territory to defend, the bands of "bachelors" do not form a hierarchy and are almost free of aggressive behavior; peaceful coexistence seems to be the rule. When, for any reason, an adult male is driven from his breeding territory, he joins the bachelors and conforms to their code of nonbelligerence.

By contrast, a male that has established a territory will make a vigorous defense of his domain, which is usually several hundred yards in diameter. Bontebok maintain their territories throughout the year, and the resident male will confront any other male that attempts to enter. Meanwhile he attracts as many females as he can and tries to keep them within his territory. During the breeding season, the buck courts the females using a series of special postures and displays—for example, approaching them with his neck and head stretched out parallel with the ground, holding his tail up and curled over his back, and lowering his ears.

491

ABOVE **A brindled gnu quenches its thirst at a waterhole before rejoining the comparative safety of the herd; on the open plain an isolated animal runs a serious risk of attack from the ever-watchful lions, cheetahs and hyenas. The collective vigilance of the herd ensures that most predators are spotted as they approach, and herds may form defensive circles to keep enemies at bay.**

FAR RIGHT (TOP) **With the onset of the first rains and the emergence of fresh grass, large herds of brindled gnu split up into smaller bands to search for food on their own.**
FAR RIGHT (BOTTOM) **Two blue cranes glide over a resting herd of brindled gnu. Gnus are inactive during the heat of the day, searching for food in the cooler hours of the early morning and late afternoon.**

will follow her, assume the courtship posture and sniff her. Normally this is sufficient to make her rejoin the rest of the group. Vigilance on the part of the male is essential, for the female bontebok goes into heat for a very brief period—just 24 hours—and during this time she may choose to mate with any male she encounters. A territorial male must keep watch on his harem at all times to stop the females from straying into rival territories.

Male bontebok normally acquire their territories at five or six years of age, marking them out in various ways: with heaps of excrement and by rubbing the secretion from their facial scent glands onto twigs and plant stems. They also leave visual marks by kneeling down on their forelegs and vigorously scraping up the turf with their horns. If another male is attracted by the presence of females and attempts to enter a territory, the resident male will follow the intruder threateningly and force him to leave the area.

Relations between males holding adjacent territories consist of a daily series of threats and displays of strength and horn length. At least 30 different patterns of territorial behavior have been recorded, and each sequence of threats and displays continues for about 10 minutes.

The restless gnus

Gnus (or wildebeest) are herbivores that live on the great plains of Africa. In the 19th century they were described as "animals having the forequarters of an ox, the hindquarters of an antelope, and the tail of a horse." Though a little on the simplistic side, this definition well describes the striking features of the animal, which stands over 3 ft. high at the shoulder and weighs up to 460 lbs. in the male.

Both sexes have strong curved horns, especially the males. The head is cattle-like in shape, thick-set, on a short neck. The legs, however, are slender and agile, while the tail is long and similar to that of a horse.

There are two species of gnu: the white-tailed gnu (or black wildebeest) of South Africa; and the brindled gnu (or blue wildebeest), which lives from northern South Africa to southern Kenya. There are five subspecies of the brindled gnu—three have black beards, and two subspecies in Kenya and Tanzania have white beards. The brindled gnu is a highly gregarious animal: lone individuals and small groups are comparatively rare.

A female receiving a male's attention may allow the male to approach and sniff her genital area to see if she is in heat. She then bounds away, stopping after a few yards. In one day, an adult male may carry out several such checks on the females already in his harem, even checking females that are still sexually immature. Although the rutting season is limited to three months, this behavior continues throughout the year. This suggests that the real purpose of his continual checking is to assert his dominance over the females. Such a view is supported by the observation that if a female attempts to leave the harem, the male

ABOVE Two male brindled gnus meet headlong, grappling on their knees as their buffalo-like horns clash to settle a dispute. Being nomadic animals, brindled gnus do not establish carefully marked, semipermanent territories like other grazing antelopes. Instead, a male stakes out a temporary territory on a small patch of ground wherever his harem of females happens to be, and continues to defend the harem against rival males when the herd moves off on the migration trail.

Mighty migrations

Sometimes, during their seasonal migrations, herds of tens of thousands may be encountered, moving in search of fresh pastures. In Tanzania's Serengeti National Park these great migrations take place twice a year. The gnus travel the whole length of the park so that they are in the southeastern sector during the rainy season and in the northwestern area during the dry season. Although they are protected in the southeastern sector, where poaching is relatively uncommon, they are slaughtered in the north, where part of their migratory route lies outside the park and is unsupervised.

During their migrations the gnu sometimes cover great distances—as much as 1000 miles—and they often walk in single file. Like many other animals of the African grasslands, brindled gnu scatter during the rainy season, when water and fresh grass are plentiful.

Once the dry season returns, the gnus come together again in large groups centered on wells or watercourses. The animals have to drink frequently and sometimes make daily journeys of over 30 miles to find water.

Temporary harems

During the rut, adult males control groups of females that may number from two or three up to more than a hundred. In some cases, very large groups of females are split up between two or three males. Each adult male defends an area of ground around his group of females, but these territories are strictly temporary affairs, for the nomadic herds rarely stay in one place for more than 12 hours. At the end of the mating season, the family groups gradually break up and the animals mingle to form large herds with no obvious social structure. Gnus are noisy animals and often emit loud snorts, grunts and squeals, especially when the animals are angry or alarmed.

The young are usually born in the morning, within the safety of the herd, and they are able to stand after only five or six minutes. By late afternoon they can already move and run. Many young gnus fall prey to predators such as lions, hyenas and Cape hunting dogs, or they die from rinderpest, a disease that attacks calves lacking immunity.

Merciless slaughter

The white-tailed gnu was once common and widespread throughout southern Africa, but about halfway through the 19th century it became a target for the guns of the Boer farmers, who killed it to provide meat for their farmworkers and to sell the hide for belts, bags and other fashion accessories. Unrestrained hunting continued for over a hundred years, reaching a peak in about 1870, when hunters carried out wholesale slaughters in order to cash in on the lucrative trade in hides, which were sold on the European market. Only 600 gnus escaped this butchery, thanks to the efforts of two Boer landowners who wanted to save the species and preserved breeding herds on their land.

The white-tailed gnu now lives under strict protection in South African parks and nature reserves, and it is no longer considered an endangered species. It differs from the brindled gnu in having a straight back, a black or dark brown coat, a white tail and a curious brush-like area of stiff hair on its face. Its horns descend forward before pointing upward (those of the brindled gnu curve down and to the side before turning upward and inward).

Roan and sable antelopes

The horse-like antelopes, such as the roan and sable, are the most impressive of the African grazing antelopes. They are large animals, weighing from 140 to 650 lbs., with massive, muscular necks; their withers are significantly higher than their hindquarters, and they have long, slender legs. Their heads are long and big and bear long, imposing backward-pointing horns with a spiral or ring pattern of ridges.

Both males and females have horns, and although the males are slightly larger and may have a different coat color (as in the sable antelope), the sexes tend to be very similar. Females have four teats, like cattle, and

ABOVE A white-tailed gnu scratches its flanks with its big, cow-like horns to rid itself of ticks in the hide. Also known as the black wildebeest, the white-tailed gnu was almost wiped out by indiscriminate hunting in the 19th century, and it owes its survival to the strict protection that it enjoys in the game reserves of South Africa.
ABOVE LEFT Two brindled gnus (or blue wildebeest) spend the hottest hours of the day resting. Part of the gnu's grooming behavior includes rubbing its face on the ground, against a tree trunk or a partner.
PAGES 496-497 Hundreds of wildebeest fording a river in Kenya form a dramatic, tangled mass as they try to climb the far bank. On the Serengeti-Mara plain, 16 percent of deaths among gnus are due to accidental drowning and broken limbs.

males and females have scent glands on their faces or between their toes. The gestation period lasts about nine months, and usually one offspring is born at a time, weighing about 35-40 lbs. at birth.

The roan antelope is the largest of the tribe. Up to 5 ft. at the shoulder, it has a light reddish-brown coat which varies in shade according to the subspecies (there are six). Its muzzle bears a dark "mask" on a white background, and it has a stiff mane on its neck, rather like that of a zebra or wild horse. The horns are long and curved back, and those of a mature male may be 3 feet or more long. The tail is brownish-black and tasseled toward the tip.

TOP Interrupted while grazing in a woodland clearing, a buck roan antelope keeps a wary eye out for intruders. Roan are the biggest of the grazing antelopes, and will not hesitate to turn and fight with an attacker.

ABOVE A pair of male sable antelopes sprint for cover across the mudflats. Sable antelopes usually establish their territories in woodland, but like their relatives, the roan antelopes, they are never found far from water.

Reliant on water

Roan antelopes are found throughout central and southern Africa in open or lightly wooded, grassy areas. They never stray far from a source of water, as they cannot survive for more than two to three days without a drink. During the dry season, the antelopes will visit a waterhole several times a day and, like the sable antelope, they frequently stay near water during the hottest part of the day.

Roan antelopes live in small groups of 5 to 20 females, each group dominated by one adult male. During the dry season these groups often gather together to form large herds of up to 150 animals. The herds move around a lot, so the males do not have fixed territories. They will defend themselves fiercely with their horns if attacked, and their rather aggressive nature and large size—exceeded only by the kudu and eland—give them an advantage over all other antelopes when grazing is scarce.

Black beauty

A mature male sable antelope is a magnificent animal, with a black glossy coat, long backward-curving horns which may be over 5 ft. long, and a stiff horse-like mane on its muscular neck. The females are slightly smaller with less impressive horns, and their coats are a dark rusty brown. Found in central and southeastern Africa, the sable antelope prefers more densely wooded country than the roan antelope, but it too never strays far from water. Its habits are very similar to those of the roan antelope, although the herds tend to be less nomadic during the breeding season. As a result, the males are more territorial and may defend a particular patch of land for two years or more. They sometimes fight fiercely for possession of females during the mating season. The roan antelope can also be extremely aggressive toward other species, and have even fought off attacking lions.

The sable antelope has four subspecies, including the giant sable antelope of Angola. As its name suggests, this is the most powerful of the sable antelopes. It has very long, finely ringed horns, a completely black coat (in the male), a thick black mane, a white belly and white facial markings. Its population was estimated at 2000-3000 in 1970. A reserve has been set up to protect the entire area currently inhabited by this rare and precious animal.

ABOVE As he sets about his courtship, a male gemsbok (right) gives his intended mate a foreleg kick as he pursues her, one of the many stylized gestures that grazing antelopes have evolved as a preliminary to the mating ritual.

RIGHT To avoid injuring each other with their long, potentially lethal horns, gemsbok and other oryxes have developed a ritualized style of combat. Here two male gemsbok clash foreheads (top), lock horns as a test of strength (center) and try to push each other over (bottom).

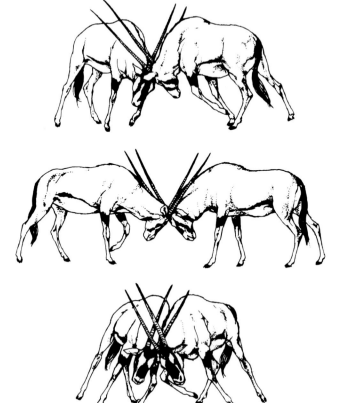

Desert antelopes

There are three species of oryx: the gemsbok, the scimitar oryx and the Arabian oryx. They are medium-sized antelopes, all sturdy, well-built animals with short manes, shoulder humps and long horns that are straight or only slightly curved with conspicuous rings. All three are adapted to life in dry environments, and the Arabian oryx is a specialist at surviving in the desert.

The gemsbok (often called the beisa oryx, fringe-eared oryx, or simply the oryx) lives in the semi-desert grasslands of the southern half of Africa. Like many of the larger antelopes, it suffered badly in the past from hunting, and the Boer farmers were responsible for almost completely eliminating the animal from the Cape Province, South Africa, during the 19th century. The name "gemsbok" is an Afrikaans name, and

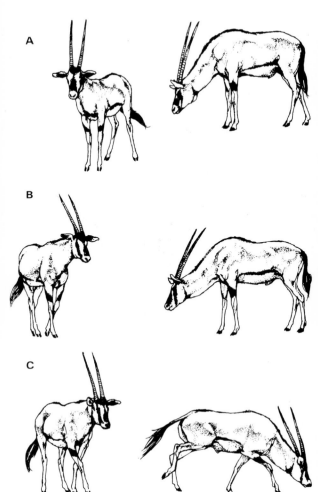

ABOVE **A group of gemsbok, a subspecies of oryx, quench their thirst at a waterhole. If water is available, oryx will drink daily, and will even dig for it if a drought is severe.**
RIGHT **A gemsbok displays its superior status: the dominant animal (left)** blocks the path of a subordinate by standing with its back to it (A); the subordinate (right) lowers its head and rakes its horns back (B); the subordinate animal retreats, with every line of its body indicating total submission (C).

actually means "chamois," although the animal bears very little resemblance to the chamois of Europe and Asia Minor. It has a fawn coat with a white belly, a black mane and tail and black and white markings on its head, flanks, rump and legs. Its horns are straight and may grow to over 3 feet long.

The gemsbok feeds mainly on grass supplemented by fruits, tubers and succulent plants that grow on the ground. The animal uses its broad front hooves to dig up these plants, which provide it with both food and water. The gemsbok also has a number of physiological features that help it to conserve water, and it can survive for several weeks without a drink in arid, desert areas. Gemsbok tend to live in herds that forage together in tight units. These contain about 30 or more animals, which include several adult males as well as females and their young.

Tamed by the Ancients

The closely related scimitar oryx was once found throughout northern Africa, but today it is restricted to a narrow strip of thinly vegetated semi-desert grasslands along the southern margins of the Sahara. Also known as the white oryx, it has a pale coat with a white head bearing pale brownish markings, a reddish-brown neck and breast, and curving, scimitar-shaped horns that grow up to 4 ft. long.

Easily tamed, the scimitar oryx was reared by the ancient Egyptians as a domestic animal. Hunting has always taken its toll, and today it is still under pressure from hunting. Its habitat is also under threat, owing to irrigation schemes designed to make these arid areas suitable for agriculture and cattle ranching. The situation of the scimitar oryx is becoming increasingly critical, and it is now considered to be in imminent

ABOVE Protected from poachers by a stout fence, two Arabian oryx stretch their legs within the confines of a wildlife park. Such parks were to prove the salvation of the species when it became extinct in the wild in 1972. With the help of a careful management program, the surviving oryx were encouraged to breed and flourish in captivity, and 10 years after their disappearance from the deserts of the Middle East, a free-living herd was reintroduced to central Oman.

danger of extinction in the wild. In 1986 a herd of scimitar oryx was established in Tunisia as part of a reintroduction program.

Nomadic creatures

Scimitar oryx are nomadic animals that wander over large areas of arid country in search of food. Their diet ranges from grass and leaves to the shoots of trees and bushes. In the desert they eat thick-leaved plants, succulent roots, wild melons, tubers and onions. They also have the same ability as the gemsbok to survive for long periods without water, and they usually rest at the hottest times of day. They live in groups of about 20 animals dominated by one adult male, but often containing several subordinate males. Fights between the males within a group are almost unknown, and instead the hierarchy is preserved by a complex system of display and highly ritualized combat.

THE ARABIAN ORYX

At one time, the Arabian oryx lived throughout the Arabian Peninsula as far west as Sinai and north up to southern Syria, but by 1972 it had been hunted to extinction in its natural habitat. It was the introduction of firearms that marked the beginning of the end for this beautiful antelope. The advent of motor vehicles then speeded up its decline, allowing the hunters to penetrate formerly inaccessible parts of the desert. Using automatic weapons, jeeps specially equipped for the desert, and even light planes, the trophy hunters finally succeeded in tracking down the very last wild individuals and wiping them out.

Ten years before the last of the original wild Arabian oryx were killed, the Flora and Fauna Preservation Society had seen the danger and launched Operation Oryx, a campaign dedicated to the preservation of this species. Under the direction of the Society, a herd was built up in captivity in the USA, and released into the wild in central Oman in 1982. Those animals should survive if, this time, the poachers' guns can be kept at bay.

501

ABOVE **Hunted to near-extinction for its magnificent spiral horns, the addax is an animal specialized for life in the desert. It seems to be able** **to detect where the next rainstorm is likely to occur, enabling it to be ready on the spot when the rain brings about a brief burst of vegetation.**

An oryx of the sands

The most specialized of the oryxes is the Arabian oryx, a relatively small antelope that rarely weighs more than 165 lbs. Its coat is almost pure white with dark chocolate-brown legs, a black tuft at the end of its tail and dark brown markings on its head. Its hooves are more rounded and larger than those of the other oryxes, and their shape probably represents an adaptation to sandy surfaces, helping to prevent the animals from sinking into the sand at each step.

The Arabian oryx lives around the fringes of deserts and can even survive life in the desert interior—at least during the cooler months—since it is well adapted to extremes of drought and temperature. Its white coat reflects the rays of the sun but is thick enough to keep the oryx warm at night. It has the stamina to walk for hours in search of food or water, and like the other oryxes it rarely needs to drink.

A desert survivor

Once found throughout the Sahara, the addax was all but wiped out by indiscriminate hunting and now survives in scattered populations in the least hospitable parts of the desert, far from water. Like the Arabian oryx, the addax has a white coat to reflect the heat of the sun, and its hooves are even broader to prevent it from sinking into the sand in dune country.

GRAZING ANTELOPES
CLASSIFICATION: 2

Hippotragini

Six species make up the tribe Hippotragini, the horse-like antelopes. They are animals of dry country and lightly wooded savannah, living in Africa and the Arabian Peninsula. Medium-sized to large antelopes, they have thick necks and long, ringed horns, present in both sexes.

The genus *Hippotragus* contains the largest species. The roan antelope or horse antelope, *H. equinus*, ranges from West Africa east to Ethiopia, and south to central and southern Africa. The sable antelope, *H. niger*, occurs in East, southern and southwest Africa.

Three species belong to the genus *Oryx*—all three have long, impressive horns. The gemsbok,

O. gazella, inhabits the semi-desert grasslands of East, southern and southwest Africa. It has five subspecies, including the Beisa oryx of Ethiopia and Somalia and the fringe-eared oryx of Kenya and Tanzania. The scimitar oryx, *O. dammah,* survives in a narrow belt of land running along the southern Sahara from Mauritania east to the Red Sea. The Arabian oryx, *O. leucoryx,* became extinct in the wild in the 1970s, but captive animals have been reintroduced into Oman in the Arabian Peninsula.

The genus *Addax* has just one species—the addax, *A. nasomaculatus*, which used to be found throughout the Sahara but now has a scattered distribution in Algeria, Mauritania, Mali, Niger, Chad and the Sudan.

HOOVES ON THE WATER MARGINS

Splashing through reedbeds or munching on marshy plants, the "water antelopes" rarely stray from their aquatic world; the mountain reedbuck, however, prefers high-altitude grasslands

Bohor reedbuck

Common waterbuck

Lechwe

Nile lechwe

Reedbucks, waterbuck and their close relatives form the largest tribe of grazing antelopes. Many of the species are associated with wetlands, and some spend much of their time feeding on semi-aquatic vegetation in swamps, lake margins and rivers. These members of the tribe are often referred to as the "water antelopes." Others, such as the reedbucks, are found on grassy plains. In all the species only the males have horns.

At home in swamps

The waterbuck is the most aquatic of the water antelopes. It nearly always lives in swampy habitats and is well adapted to its environment; its long, rough coat, with a characteristic mane along the neck, is continually lubricated by sebaceous (fatty) glands to make it waterproof. The male has long horns that curve elegantly upward and forward. The size of a large European deer, the waterbuck can weigh up to 550 lbs. and is found in many parts of Africa south of the Sahara (especially in the east) in savannah and woodland wherever there is a good water supply. It does not live in the equatorial rain forests. There are 13 subspecies, divided into two groups: the common waterbuck has four subspecies, all found in eastern and southern Africa, while the defassa waterbuck of western Africa has nine subspecies.

The young males leave their mothers at about nine months old and join with other young males to form a single-sex, subadult bachelor herd. These groups vary in number from two to 35, with an average membership of five, but they are not stable and the numbers fluctuate from month to month.

"Satellite" defenders

Adult males establish territories that normally include a stretch of swamp or riverside vegetation. They will not tolerate the bachelors on their territories, and if the young males attempt to cross the boundary they are driven off. In some areas, such as East Africa, where the population densities are high, one or two subordinates or "satellite" males may join a territorial male and help him defend the territory, but they rarely attempt to mate with any females.

The advantage the subordinates derive from this is unclear, but the evidence suggests that they may inherit the dominant male's territory if he is displaced, whereas free-ranging bachelor males are in a much

ABOVE Ears pricked, a young waterbuck copies its mother as she checks for signs of danger. The shaggy coats of these big antelopes are greased with an oily secretion that repels water — a valuable adaptation for an animal that spends much of its time wading in swamps, lakes and rivers, eating water plants and bankside vegetation.
PAGE 503 A group of male defassa waterbuck make for cover after they have been disturbed while out in the open.
PAGES 506-507 A male waterbuck shows his interest in a female by nudging her gently from behind with his foreleg.

weaker position. The subordinate males may also gain access to better grazing along the water's edge, which is normally claimed by the dominant males as part of their territory.

Sexual maturity in waterbuck is reached at about 13 months in females and 14 months in males, although males cannot usually dominate a territory until they are six to seven years old. The animal remains at its peak until about 10 years of age. The cost of maintaining a territory is, however, high in terms of energy expenditure, and a male is generally forced to abdicate in favor of a younger rival when he reaches the age of about 11 years. Having lost his territory, the older male does not return to the bachelor herd and generally goes off on his own.

ABOVE Two male waterbuck greet each other. In areas where the waterbuck population is very high, two or three males may join forces to defend a territory, although the dominant partner will claim any females that enter the territory.
BELOW The map shows the distribution of the reedbuck and some of its close relatives.

Territorial mayhem

The extent of each territory varies according to the age of the occupying male, but if he dies, his neighbors do not always extend their own dominions. They may occasionally range over this "free" zone, but they do not defend it or prevent another male from occupying it. If several territories are abandoned at the same time, however, the situation can slide into a free-for-all, and the territorial boundary map may be completely redrawn.

Territorial size is also related to population density. In some areas of eastern Africa, the waterbuck population has been observed to fluctuate from two or three males and females per 250 acres to peaks of 18 animals or more (with an average of 10 animals per 250 acres). Male territories measured some 200 acres at higher densities and 500 acres at lower densities. These figures show that there is no perfect territorial size.

Waterbuck are dependent on water, needing to drink at least once a day to balance a high-protein diet; in the dry season they eat much aquatic vegetation. Since the territorial system is strong, difficulties can arise when a male needs to cross another male's territory to reach water. Since not all the territories are bounded by water, this is a common occurrence and

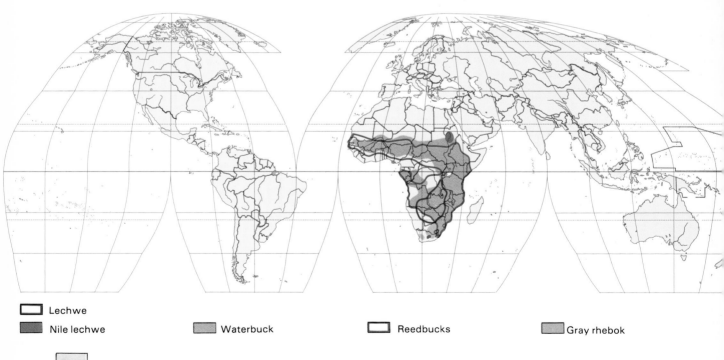

Lechwe

Nile lechwe · Waterbuck · Reedbucks · Gray rhebok

waterbuck have developed a range of behavioral adaptations to cope with this problem. The area around the water may become neutral ground where males control their aggressive instincts, and "right of passage" across another animal's territory is permitted —but only if the intruding male adopts an obviously submissive attitude as he crosses over.

Females are not territorial. They occupy loosely defined grazing areas in small herds of three to six individuals. The areas are shared with other groups of females. Each female grazing range may overlap several male territories, and a resident male will mate with any receptive females that enter his domain.

Butted into independence

Remarkably, the females operate a system whereby young but self-sufficient animals are expelled from the group. At eight to nine months old, the young animals are gently head-butted until they leave the maternal group. Although the aggression is understated, the young animals are made aware that they are no longer wanted, and before long they take off on their own. The young males form bachelor herds, while the young females form similar groups and roam around freely until they reach maturity—an event marked by their first mating at the age of about three years.

The site chosen for the first birth becomes a fixed point of reference for the young mother in the future. Her subsequent sexual activities are always conducted in this area, and she will often give birth in the same place for several years running.

Animals with a social structure based on a hierarchy often develop an extensive range of warning gestures, and waterbuck are no exception. The threat display involves dashes in the direction of the intruder with lowered head—though without pointing the horns at the enemy—while the male indicates his dominance by raising his head and displaying his throat.

Courtship kicks

The courtship ritual is often preceded by a series of foreleg kicks. These are not violent, for the male merely touches the hind legs of the female with his straight front leg. The two partners place themselves head to tail, and the male rests on the female while she makes biting motions. A true mating season does not exist in most areas, and peaks of seasonal sexual activity are noted only in certain areas.

ABOVE **A young male defassa waterbuck canters across the savannah of central Africa past a huge flock of pelicans. Waterbuck are rarely found far from permanent water; those in captivity** **need to drink a lot to balance a diet that is unusually rich in vegetable protein. By contrast, other grazing antelopes, such as the oryx, can go for days without drinking.**

The kob

Kobs are smaller than waterbuck, weighing up to 265 lbs. at most. They have no manes, their coats are shorter and their horns are curved into a sinuous, open S-shape. The 10 subspecies are found in Africa south of the Sahara, from Senegal in the west through central Africa to Uganda in the east. The best-known subspecies are the white-eared kob and the Uganda kob. The territorial behavior of the Uganda kob includes the unusual feature of mating "arenas."

Research in the Ugandan Toro Game Reserve in the 1950s showed that the population of 18,000 kobs was organized into 13 subunits, each containing 1000 to 1500 individuals. These subunits constantly revolved around a series of individual, fixed mating grounds termed "arenas." Each arena consisted of a set of 30 or 40 circular male territories, each 25 to 60 ft. in diameter and vigorously defended, while the whole complex measured about 650 ft. across.

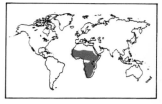

The three tribes of grazing antelopes are confined to Africa and the Arabian Peninsula.

The center of each territory and the arena in general was free from grass and vegetation, with considerable stretches of exposed, bare soil—although the boundaries between individual territories were marked by grass. In the Toro Game Reserve, the female was free to decide when mating would take place. The preliminaries might occur in individual territories outside the arena, but it was only when one or more females entered the arena that true courtship and mating took place.

A considerate suitor

When a female Uganda kob approaches or enters the breeding ground or arena, the male becomes tense. His neck stiffens, he arches his head backwards with horns slanted and chin up, and he walks with a measured, stiff-legged gait. He sniffs her urine to gauge whether she is in heat; if she is, he will approach her cautiously and assess her receptiveness by touching her gently with his foreleg. The female usually moves away and the male follows her, continuing to court her. After mating has taken place, the kob, uniquely among the ungulates, continues to attend his companion. He stays with her, touching her with his leg and uttering a whistle that may be taken up in chorus by other animals in the area.

ABOVE A pair of defassa waterbuck size each other up with a view to mating. Dominant males tend to claim all the best grazing land—which is often alongside lakes and rivers —as part of their territory. While females are allowed free access, the younger males often have to make do with poorer pasture. If a young male has to cross a dominant male's territory to reach water, it must adopt an elaborately submissive attitude to ward off attack.

GRAZING ANTELOPES
CLASSIFICATION: 3

Reduncini

Nine species of antelopes make up the tribe Reduncini, part of the subfamily Hippotraginae. Medium-sized to large antelopes, they inhabit grasslands and wetlands over much of sub-Saharan Africa. Horns are present only in the males.

The three species of reedbucks belong to the genus *Reduna*. The southern or common reedbuck, *R. arundinum*, occurs in southern Africa, reaching north into Tanzania and west to Angola. The mountain reedbuck, *R. fulvorufula*, has isolated populations in Cameroon, Ethiopia, and southern Africa, while the bohor reedbuck, *R. redunca*, occurs in a belt below the Sahara from Senegal to Ethiopia, reaching south as far as Tanzania. The gray or Vaal rhebok, *Pelea capreolus*, is the only member of its genus. It inhabits grassy upland plateaus in southern Africa.

The remaining five species belong to the genus *Kobus*. The kob, *K. kob*, ranges from Gambia east to Ethiopia and Uganda, preferring land close to permanent water, while the puku, *K. vardoni*, ranges from Tanzania into southern Africa. Both the lechwe, *K. leche*, of southern Africa, and the Nile or Mrs. Gray's lechwe, *K. megaceros*, of Sudan and western Ethiopia, live in wetland habitats. The waterbuck, *K. ellipsiprymnus*, is the largest member of its tribe and ranges over most of sub-Saharan Africa. Its 13 subspecies are divided into two taxonomic groups: four belong to the common waterbuck group, nine to the defassa waterbuck group.

When rival male kobs meet, they face one another at a short distance but turn their heads aside through an angle of 90 or even 180 degrees. The significance of this posture is not clear. Some zoologists have suggested that the adversaries are showing their readiness to abandon the confrontation and withdraw.

The kob has a highly selective system of reproduction which favors the strongest males, since only the most capable are able to hold onto territories in the heart of the arena. These territories obviously have to be left unoccupied when the owners go off in search of food or water, and they may be invaded by other males, but a strong male is usually able to reassert his rights. It is important for the male to do this, for the females show a clear preference for certain territories and thus certain males. Out of 30 to 40 mature males, only three or four will mate frequently. Males occupying outlying territories, even if they are still nominally within the arena, have little chance of success.

As well as discriminating in favor of stronger males, the arena system encourages a high rate of reproduction. The females may go into heat at any time of year, and the arenas are permanent. As a result, mating, pregnancy and birth all follow a continuous cycle which is faster than the annual cycle followed by other species.

ABOVE Although almost as big as its mother, a young kob still relies on her for an occasional drink. Kobs have a high reproduction rate, giving birth at intervals of about 280 days, instead of the usual 365 days. After a gestation of eight months, the female gives birth to a single offspring which she nurses for about 40 days. She then mates again. Kobs have a high infant mortality—more than 45 percent of young kobs die in their first year, most killed by hyenas and lions.

Inbreeding risk

The normal female cycle of the Uganda kob follows this pattern: after weaning at six to seven months, the young female is driven away by her mother, and while she may remain within the herd, she will often form a group with other young females. At 13 months old she goes into heat for the first time, and visits the arena to mate with the male of her choice.

An inevitable consequence of the favoritism of the females is that three or four males are responsible for all the coupling on any one day. The risk of inbreeding would seem considerable. But in practice the problem does not arise because the females spend less than 15 percent of their time in the vicinity of any one arena, and are quite likely to visit different arenas for subsequent matings. The movements of subadults also ensures genetic interchange between the herds.

Mating exhaustion

The turnover of breeding males is high. It is true that three or four males mate with all the females on a daily basis, but by so doing they become exhausted quickly. Continual fights with other males to maintain their status make their task even harder. A male can rarely last more than 10 days in the mating arena—some males may give up after only one day—before retreating to muster strength back in the bachelor herd. Since there may be 500 females in breeding condition in the area, many remain unmated.

Naturally, the most favored males will be the most overworked, and therefore the first to tire; their role is then inherited by other males, and it has been calculated that about 50 adult males may take turns mating with all the females who enter the arena.

Life in the swamps

The lechwe of southeastern Zaire, Zambia and Botswana is similar to the kob in appearance, but is more compact and better adapted to life in the swamps and seasonally flooded river valleys. Its coat is longer and thicker than the kob's, and its long, spreading hooves provide a bigger surface area for walking or running over soft, waterlogged ground. Its lyre-shaped horns are also longer, thinner and more impressive. There are three subspecies, including the black lechwe (or bangweolo) of northern Zambia,

ABOVE LEFT A buck Uganda kob splashes through the water up to its thighs to escape predators such as leopards or lions.
ABOVE A pair of male kobs lower their horns as they square up for a fight (top). A dominant female holds her head high as she steps confidently through her territory (bottom).
FAR RIGHT Heads together, dueling male kobs fight for a scrap of ground in an attempt to enlarge their territories and so improve their chances of mating.

which is characterized by the dark, blackish coat of the male. The other subspecies—the red lechwe (or Zambezi lechwe) and the Kafue lechwe (or brown lechwe)—are bright chestnut-red with white underparts.

The lechwe has become a specialist at feeding on the lush grasses that sprout in response to seasonal inundation of the low-lying floodplains of central southern Africa. One of the most aquatic of the grazing antelopes, it can swim well and travel quickly over sticky mud, often jumping along with its legs together. Gathering in herds of several hundred, the lechwe graze in the shallow floodwaters up to 20 in. deep, cropping the grass that grows just beneath the surface. But when heavy rainfall raises the water level, they may venture into much deeper water to gather the submerged vegetation. Mass migrations often occur in response to localized flooding.

Like the kob, lechwe use traditional breeding grounds for mating. Usually some 1650 ft. across, the arenas may contain the tightly packed breeding

territories of up to 100 males, which compete for the attention of the females in the same way as male kobs. In the case of the lechwe, though, there is a better-defined annual breeding season that lasts for only a few weeks; after the rut is over the arenas are abandoned, so they never acquire the well-beaten appearance of kob breeding grounds, which may be in continuous use for 30 years or more. During mating the male lechwe scent-marks the female by spraying his beard and mane with urine and rubbing them on her rump and muzzle.

Aggressive encounters between male lechwe are preceded by threats, during which the combatants lower their heads to present their long, pointed horns. If the situation degenerates into a fight, the animals will grapple with horns interlocked.

ABOVE **Silhouetted against the glittering waters of a flooded river, two male lechwe make a dash for cover. The graceful lechwe is a specialist at feeding on the floodplains of African rivers, taking advantage of the lush vegetation that springs up beneath the fertile, silt-laden floodwaters. They often eat the grass while it is still submerged, wading up to their necks in the water if necessary.**

A Nile habitat

The Nile lechwe is a very similar species, but the male has a dark brown coat with a white shoulder patch. The female is brown all over. The animal is also known as Mrs. Gray's lechwe, after the wife of the naturalist who first classified it. Its other name is a reference to its habitat: the swamps of the White Nile and its tributaries in the border country between southwestern Ethiopia and the southern Sudan. Here it lives in much the same way as other lechwe, in mixed herds of 50 or more animals. The closely related puku has a similar way of life, but a much more widespread distribution throughout central southern Africa. Its horns are shorter, thicker and straighter, and it has a golden yellow coat.

The specialized life-style of the lechwes means that they rarely stray far from water, and the movements of the herds can be predicted with some accuracy. This lays them wide open to organized hunting, and the lechwe drive is part of the traditional way of life in many areas. Normally such traditional hunting barely dents wild populations, but the introduction of firearms has enabled the hunters to kill several thousand animals at a time. The lechwe are also threatened by loss of habitat; in an area that has become notorious for crop failures and widespread famine, building dams and setting up drainage schemes to free the land for agriculture have become an important priority. Inevitably, such schemes will reduce the amount of seasonally flooded land available to the lechwe, which is probably too specialized to change its way of life.

The whistling reedbucks

There are three species of reedbuck, ranging from the size of a small roe deer to that of a fallow deer. The males and females of each species are much the same size, but the females do not have horns. They are athletic, graceful animals and often move in a series of high, bouncing leaps. The habit extends to the mating ritual, for during courtship the male stands behind the female and, after touching her with his front leg, performs a series of rhythmic jumps. One peculiar feature of all the reedbucks is that they communicate in a series of whistles.

The two larger species—the bohor reedbuck and southern (or common) reedbuck—live in the lowlands on savannah grasslands and among reedbeds, and like many of the other water antelopes they are rarely found far from water. In contrast, the smaller mountain reedbuck lives on mountain grasslands up to 16,500 ft. above sea level, and is divided into three distinct populations living in three isolated highland regions in Cameroon, Ethiopia, South Africa and southern Mozambique.

Hidden in grass

The bohor reedbuck is found on the savannahs of northern central Africa, from Senegal to the Sudan and Ethiopia, and south into western Kenya and Tanzania. It always lives near rivers where vegetation is plentiful, and spends much of the day in the cover of long grass or reeds, emerging at night to feed. Bohor reedbuck are not especially gregarious. Each adult female tends to live alone or with her offspring, although she may be accompanied by a young female —her offspring of the previous year. Occasionally groups of up to four females and their young occupy the same area, and where the cover is poor they associate in small herds, although this is unusual for the species as a whole.

Normally each female settles within a territory of about 50-75 acres whose boundaries intersect with those of other females.

ABOVE Horns interlocked, two male lechwe dispute each other's right to a patch of breeding territory. Both lechwe and kobs use traditional breeding "arenas"—clusters of small territories. The males vie with each other for the most favorably sited territory, for only the best-placed get the chance to mate. When the dominant males retire to gather their strength, they are replaced by young hopefuls that try to get a spot likely to attract the females.

A jealous guardian

Adult males occupy territories extending over 125-150 acres, and in regions with a high reedbuck population the area of influence of each male encompasses the territories of several females. He guards the females jealously, and rarely displays aggressive behavior except in their presence. In effect, this amounts to a type of harem, and although the females do not forage as a group, they are much more sociable than females defended by different males. It is noticeable, for example, that the territories of females associated with the same male tend to overlap considerably.

The mountain reedbuck appears to be more sociable than the bohor reedbuck; the females and their young live in small groups of two to six. These groups may divide their time between several males or remain faithful to one, and they rarely travel far.

The southern (or common) reedbuck is another lowland animal, found on the southern savannahs of Zaire, Zambia, Zimbabwe, Malawi, Tanzania, Angola and in Natal and the Transvaal in South Africa. Unusually, these animals will hold territories as pairs, at least at certain times of year. The parental care also differs in a number of ways. The young female leaves her mother when the following year's offspring are born, whereas a young male will stay with her until he is two or two-and-a-half-years old.

The opposite is true of the bohor reedbuck, in which the young male is driven from the adult territory very early. Young males driven out by their mothers join up to form bands of subadult or nonterritorial males, but since they are not very gregarious by nature, these bands tend to be quite small. Although they are nonterritorial, their home ranges are well defined and limited by the territorial boundaries of other males.

Mountain survivor

The gray (or Vaal) rhebok of southern Africa is about the same size as the European roebuck, with a sloping back that is lower at the withers. Its horns are slightly different from those of the reedbucks: they are short, simple and straight instead of S-shaped and forward-curving. It lives in mountain areas with plenty of vegetation and bushes, in small herds of less than 10 individuals, and its social organization appears to be very similar to that of the mountain reedbuck.

TOP A startled waterbuck uses its powerful hind legs to leap clear of the cover. When hunted by predators, it may take to water and defend itself with its horns and hooves.
ABOVE A pair of southern reedbuck glance up while foraging amid a lakeside reedbed. Reedbuck are unusual because they sometimes occupy a territory as a pair. They feed mainly on grasses and small herbaceous plants, and in farming areas where they will sometimes raid the fields to feast on the young crops.

GRACE AND AGILITY

Nimble antelopes of Africa and Asia, the gazelles rely on speed and sudden leaps to outrun or outwit their many enemies

Gerenuk

Mountain gazelle

Dama gazelle

Dibatag

Mongolian gazelle

Springbok

Slender, long-limbed and graceful, the gazelles are among the most elegant of the hoofed mammals. There are 18 species, ranging over southern, northern and eastern Africa, through Arabia and central Asia to eastern China.

Gazelles are small antelopes with well-developed scent glands. In most species both the sexes possess horns, although those of the females tend to be shorter and slimmer than those of the males. Gazelles generally live in hot, dry environments, feeding on a mixed diet of grass, green plants and leaves.

A pale Saharan

A scarce native of the Sahara Desert, the dama gazelle is the largest of the gazelles, with a long neck and particularly long legs. It has a distinctive color scheme—reddish-brown fur extends from its neck to its back, but its head and the rest of its body are white. It also has a white collar-like mark near its throat. The proportions of brown and white in its coat vary considerably between local populations, and some animals are pale over most of their back. Its short horns are also distinctive, for they are sharply raked back, giving them an almost streamlined appearance.

Dama gazelles browse on the bushes and woody plants that grow in parts of the Sahara, and they can survive for several days without drinking. Nevertheless, they are not as well adapted to drought conditions as many other gazelles, and in the dry season they have to migrate into less arid terrain.

Soemmerring's gazelle ranges from the Horn of Africa north to Sudan and is still quite common, particularly in Ethiopia. It prefers hilly areas with plenty of bushes and acacias, but it also lives on open plains. The coloring of its coat is typical of that of most gazelles—the upperparts are fawn, the underparts pale, and the face has well-defined markings. A pale patch covers each side of the animal's head, crossed by a dark band running from the muzzle to the eye. It has a short neck, and its horns curve in toward each other at the tips.

Resistant to the sun

The Grant's gazelle normally lives on open grassy plains with varying amounts of scrub and trees, avoiding areas with high grass. In the northern part of its range it also lives in areas of semi-desert. It eats leaves and some grass, and can go for long periods

TOP Although the dama gazelle spends much of the year in remote parts of the Sahara Desert, it has suffered from relentless hunting and is close to extinction in many places.
ABOVE A family group of Grant's gazelle—a male, two females and a calf— in their typical habitat. The slender horns of the females stand in marked contrast to the large, thick, strongly ridged horns of the male (right).
PAGE 517 A female gerenuk balances on her hind legs to browse on the topmost leaves.

519

without drinking. On the African plains, the Grant's gazelle is one of the few hoofed mammals able to stay out in the sun during the hottest part of the day without seeking the shade of trees or taking to water.

There are numerous subspecies of Grant's gazelle that can be distinguished by their coat color and horn shape. Some have lyre-shaped horns, while others have horns that flare apart for the upper two-thirds of their length. They vary from fawn to reddish-brown on their upperparts and are pale below, with a black rump stripe and a black tip to the tail.

During the dry season, Grant's gazelles congregate in large groups of 40-400 in areas where there is still vegetation. At other times it is rare to encounter more than a few dozen individuals in the same place. The adult males are territorial, and the areas defended by each male vary from a few hundred yards to well over half a mile in diameter.

The males mark the boundaries of their territories with mounds of dung in typical bovid fashion, but they also use more energetic means, such as scattering bushes and plants with their horns. When one male intends to threaten another, he approaches stiff-legged, with his head up and horns clearly visible. The two contestants then circle one another and take up a series of threatening postures as a preliminary to combat. The displays are usually sufficient to convince one of the rivals to back down, and as a result, actual physical combat is rare. However, when both animals happen to be of equal rank and neither will back down, they have little option but to fight with horns interlocked, each pushing the other and apparently trying to twist his opponent to the ground. Occasionally, they will strike at one another's flanks and may cause quite serious wounds.

The rare mountain gazelle

The mountain gazelle, Arabian gazelle or idmi, once occurred from the Arabian Peninsula through Iran to India. Today it is very rare and restricted to a few areas where it has been fortunate enough to escape the effects of human settlement. It has a sand-colored coat, a dark band across each flank and short, almost stumpy horns.

The social organization of the mountain gazelle is typical of the gazelles as a whole. Adult males claim breeding territories, while young males join together to form single-sex bachelor herds. Females and young form separate herds but associate with the territorial males during the mating season. Most of the aggressive encounters between males take place with animals of the same age, sex and rank, and the

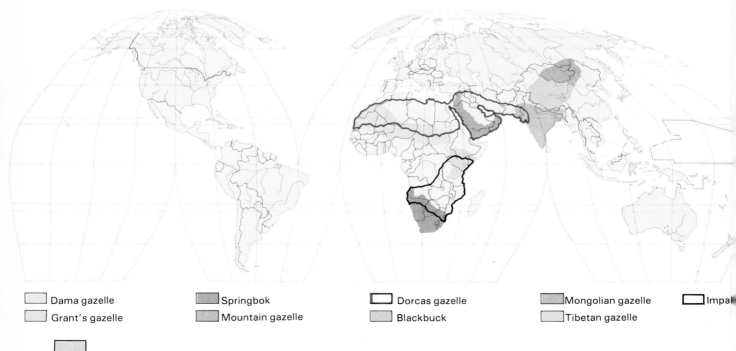

Dama gazelle   Springbok   Dorcas gazelle   Mongolian gazelle   Impal

Grant's gazelle   Mountain gazelle   Blackbuck   Tibetan gazelle

A

ABOVE **Sharing their battleground with a trio of zebras, two male Grant's gazelles interlock their horns and wrestle for their territorial rights. If one male breaks free and strikes the other's flank, he may inflict a serious wound.**
LEFT AND BELOW **Grant's gazelles often avoid a** fight by using threat displays. Here two males are shown approaching each other with horns to the fore (A), displaying with heads turned aside (B) and with heads raised and horns back (C).
FAR LEFT **The map shows the world distribution of the impala and eight species of gazelles.**

B

C

THE THOMSON'S GAZELLE
— SHARP EYES FOR SURVIVAL —

The Thomson's gazelle is one of the most familiar animals of the East African savannah. It is still numerous, particularly in the great African wildlife parks. (Some 200,000 individuals are estimated to live in the Serengeti National Park alone.) It owes its survival in such numbers to its relatively small size and unspectacular appearance, factors which make it a less attractive target for poachers and hunters. It has, therefore, escaped the persecution that has reduced great numbers of larger herbivores with more striking horns and coats, or more meat.

Although the Thomson's gazelle has escaped the unwelcome interest of man, it remains a favorite prey of many carnivores, including lions, hyenas, Cape hunting dogs and cheetahs. Jackals may attack the young gazelles, and although the mothers fiercely defend their offspring, they cannot always save them.

An eye on danger

The first warning of an approaching predator is usually visual. The gazelles' senses of hearing and smell are good, but they do not seem to play an essential role in identifying the predator, and their keen sight is almost certainly their most important aid to defense. The adult territorial males act as guardians of the herd when predators appear. They watch the intruder carefully—assessing its intentions— and keep it in sight as the rest of the herd keeps well away.

The young, subadult, non-territorial males are usually the first to raise the alarm, not because they have better sight but because they graze on the edge of the herd and have the best view of the surrounding landscape. They are also the most exposed to attack, and although they are well equipped to escape, being young and strong, young males form a large proportion of the gazelles taken by carnivores. From the point of view of the herd, this is the best outcome, for most of the young males will have little chance to breed, and they merely compete with the breeding females and their young for pasture.

Evading the jaws of death

When a gazelle is alarmed, it usually retreats at a gallop broken with long leaps; young animals often adopt a strange springing gait. Out on the savannah there is no cover to aim for, so the survival of each gazelle depends on its ability to outrun a predator and discourage it from continuing the chase. Once the enemy has broken off the attack—probably because it has found easier prey—the gazelle can slow down and stop. The distance a gazelle will run depends on the predator and its hunting techniques. It may vary from 16 feet for jackals and hyenas, which never give chase if they can avoid it, to more than half a mile in the case of Cape hunting dogs, which hunt by running down their prey until it is exhausted.

Thomson's gazelles can be aggressive toward their own kind, and disputes are frequent. They usually take place between gazelles of the same sex and age, or individuals positioned immediately above or below one another in the herd hierarchy.

RIGHT Delicate and vulnerable, a young Thomson's gazelle may be attacked by animals as small as jackals. It would be an easy target for the jackals if it were not fiercely protected by its mother. Against larger carnivores, the adults have little defense other than their sprinting speed.

BELOW LEFT A mixed group of male and female Thomson's gazelle. When the grazing is poor, vast herds containing thousands of gazelles may gather to migrate to better pastures elsewhere.

BELOW A Thomson's gazelle uses its hind feet to scratch its head, neck and breast. Grooming also involves licking and nibbling of the fur.

BELOW RIGHT Like many other animals with well-developed horns, Thomson's gazelles have a highly ritualized fighting technique. Here one male is shown threatening an intruder (A). The intruder responds with a challenge, turning his horns toward the threat (B). The two engage in a brief tussle with horns locked (C) before the intruder backs off and retreats with the victor in pursuit (D). Had there been an all-out fight, they might have risked serious injury.

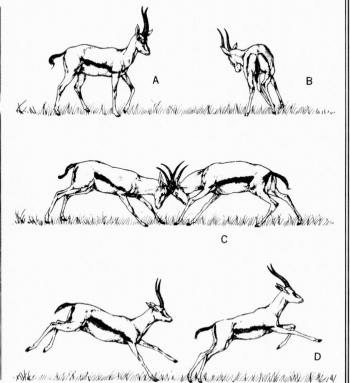

TOP A rare glimpse of a mountain gazelle. Once widespread over the Middle East and ranging as far west as India, this species has lost much of its habitat to agriculture and ranching. It is now restricted to the less hospitable parts of its former range, mainly within the Arabian Peninsula.

ABOVE Watched by guinea fowl, a male red-fronted gazelle drinks from a waterhole. Like many animals that live in arid environments, it can survive on little water and may satisfy most of its needs by taking in dew and the moisture trapped in plants.

intensity of the disputes increases with the age of the contenders. Adult males, however, generally resolve their differences without aggression by using a series of ritual threats and displays of strength. This avoids the risk of injury that is always present in real fights.

The "Tommy"

Thomson's gazelles (often popularly known as "Tommies") are small antelopes, still common in many parts of northern Tanzania and Kenya, with an isolated population in the southern Sudan. Fifteen geographic races have been described, but in all these the coat color is reddish-brown with a broad black band along each flank and a white underside. They also have distinct black and white markings on the face, similar to those of Soemmerring's gazelles. The males have long, prominently ridged horns with slight backward curves but forward-pointing tips; the females' horns are much smaller.

An animal of the open plains, Thomson's gazelle usually avoids areas of thick scrub or tall grass. It lives in small groups of 60 or more individuals that are dominated by a single adult male. These groups are very unstable—their size and composition may change from hour to hour—and thousands of animals may unite to form vast herds during the great seasonal migrations in search of richer grazing grounds.

Survival in arid regions

The dorcas gazelle is distributed through North Africa and the Near East, but it is becoming increasingly rare and is in danger of extinction throughout its range. Inhabiting semi-desert areas, it has become well adapted to survival in arid sandy or stony terrain. It is able to withstand prolonged exposure to the hot desert sun, and like Thomson's gazelle it has to drink very little, for it obtains the water necessary for its survival from succulent plants. It is a browsing animal, feeding on the foliage of acacias and other trees and shrubs, and occasionally locusts. Adult males are territorial and generally solitary.

Swollen larynx

The goitered gazelle is an Asian species, found from Iran to Mongolia. Once widespread, it has declined greatly, especially in the western parts of its range. It owes its name to a strange goiter-like swelling of the larynx (voice box) in the throat. The swellings are

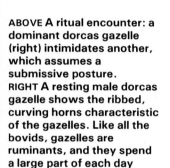

ABOVE A ritual encounter: a dominant dorcas gazelle (right) intimidates another, which assumes a submissive posture.
RIGHT A resting male dorcas gazelle shows the ribbed, curving horns characteristic of the gazelles. Like all the bovids, gazelles are ruminants, and they spend a large part of each day chewing over the food they have already gathered.
PAGES 526-527 Two male Grant's gazelles clash over a territorial dispute. Territories may stretch more than half a mile in width, their perimeters marked by dung piles, scrapings on the ground and disturbance to nearby vegetation.

developed by the males during the mating season, and they also have enlarged necks. The horns are set close together and converge at the tips. Unusually for gazelles, horns are only present in the males, although some females are equipped with two short stumps. The mating season lasts throughout the winter, but activity reaches a peak in late autumn. Twin births are frequent, and even triplets are occasionally born.

While other gazelles are well adapted to high temperatures, the goitered gazelle has acquired considerable resistance to the low temperatures and snowfall that occur in parts of its range. On occasion it may be completely buried in the snow, but will survive provided it can breathe. The animals often gather in groups at lower altitudes to avoid worse conditions higher up.

Asiatic gazelles

Three closely related species of gazelle are native to central and eastern Asia: the Tibetan gazelle or goa, the Mongolian gazelle or zeren, and Przewalski's gazelle. They are medium-sized gazelles, rarely growing taller than 25 in. at the shoulders, and all are well adapted to the harsh climate of central Asia with its extremes of heat and cold. They have short brownish summer coats that develop into thick woolly

pelts in winter. The horns, only possessed by the males, are simple ringed prongs like those of other gazelles.

The Mongolian gazelle is widespread over the barren plains of Inner and Outer Mongolia. It is slightly bulkier than the other two species, weighing up to 90 lbs., and during the mating season the males develop a throat swelling like that of the goitered gazelle. Mongolian gazelles often give birth to twins; Przewalski's and Tibetan gazelles follow the usual gazelle pattern and produce only one offspring at a time. In all three species the young are born in late spring. After calving, the two sexes separate and remain apart for most of the summer, living in single-sex groups. They come together again for the autumn mating season and then gather into large mixed groups for migration to the wintering grounds. These stand at lower heights where the climate is less severe.

The sacred blackbuck

Distributed throughout India up to the Himalayas, the blackbuck is one of the most distinctive of the gazelles, owing to the deep chocolate or black coat of the male. The dark coloration extends over its back, neck and much of its head, and contrasts sharply with its white eyepatch, chin, breast and belly. It is a handsome, square-cut animal with long, spiral horns

ABOVE Hunted for centuries for its magnificent spiral horns, the blackbuck of India is becoming increasingly rare. It has been introduced into North America, and its numbers there now probably exceed those in its native land. RIGHT Head raised in a display posture, a male blackbuck courts a female (A). When the single young is born, it spends most of the time alone, lying low for safety (B). Called by its mother (C), it runs to her and, after a brief sniff of recognition (D), it is allowed to drink (E) before returning to its hiding place.

that may grow to 30 in. in length. The females, on the other hand, are not only hornless but different in color, with light brown coats much like those of other gazelles. The blackbuck is one of the few antelopes in which the sexes have distinct coloration.

The blackbuck lives in a variety of habitats, ranging from open woodland to arid, semi-desert plains. Sadly, it is becoming increasingly rare, and the decline seems to be due mainly to uncontrolled hunting. Areas where the animal is still respected and protected as a matter of course (it is considered a sacred animal in some parts of India) are on the decrease.

Blackbuck are tolerant of heat and are able to stand the fierce tropical sun without seeking the shade of trees or rocks, although they usually rest during the hottest part of the day. They generally live in open spaces, where they feed on foliage and grasses. They need drink only infrequently—a useful attribute for an animal that spends much of its time in a dry, semi-desert environment.

If a group of blackbuck is alarmed by a predator such as a tiger or a human hunter, they will bunch up and take flight together—an impressive sight. When alarmed, one of them leaps up to six feet into the air, and the others soon copy the action. After a few such "bounces," they settle down to a regular gallop. Blackbuck run very fast and can usually outsprint greyhounds, but cheetahs can catch them and have even been trained by humans to do so.

Long-necked gazelles

Of all the gazelles, the most remarkable must be the gerenuk of eastern Africa—remarkable both for its extremely slender, delicate build and for its specialized feeding technique.

Standing on its long, slim legs, the gerenuk is one of the tallest of the gazelles, measuring over three feet at the shoulder. It also has a very long neck that supports a small head with large eyes and long pointed ears. The males have tapered, lyre-shaped horns that curl forward at the tips, but the females are hornless.

ABOVE The elongated body and superb balance of the gerenuk give it an advantage over most other gazelles, enabling it to browse in comfort on the upper leaves of bushes and small trees. Because few other animals exploit this high-level food source, the gerenuk can be selective in its choice. It generally picks only the young, succulent foliage, ignoring the old fibrous leaves. The juicy leaves provide all the water it needs, and it rarely has to drink.

GAZELLES CLASSIFICATION

The gazelles belong to the tribe Antilopini, part of the subfamily Antilopinae. They range over Africa and Asia, and are all slender in build, with long, thin legs. There are 18 species grouped into six genera.

The largest genus by far is *Gazella*, with 11 species. Seven of these are fairly small animals and include Thomson's gazelle, *G. thomsoni*, of East Africa; Speke's gazelle or dero, *G. spekei*, of Ethiopia and Somalia; the red-fronted gazelle, *G. rufifrons*, which lives on the southern semi-desert fringes of the Sahara; the edmi or Cuvier's gazelle, *G. cuvieri*, and the slender-horned gazelle, sand gazelle or rhim, *G. leptoceros*, of North Africa; the mountain gazelle, Arabian gazelle or idmi, *G. gazella*, of Arabia; and the dorcas gazelle or chinkara, *G. dorcas*, which is found in fragmented populations from North Africa through southwest Asia to India.

The four larger species in the genus are the goitered gazelle or Persian gazelle, *G. subgutturosa*, of Arabia and central Asia, Grant's gazelle, *G. granti*, of East Africa, Soemmerring's gazelle, *G. soemmerringi*, of Ethiopia, Somalia and the Sudan, and the dama gazelle, *G. dama*, of the Sahara Desert.

The genus *Procapra* includes three Asian species: the Tibetan gazelle or goa, *P. picticaudata*, the Mongolian gazelle or zeren, *P. gutturosa*, and Przewalski's gazelle, *P. przewalskii*, of China.

The other four genera in the tribe have one species each. The blackbuck, *Antilope cervicapra*, is found in India and Pakistan. The springbok (or springbuck), *Antidorcas marsupialis*, is a southern African species, while the gerenuk or Waller's gazelle, *Litocranius walleri*, occurs in East Africa, and the dibatag or Clarke's gazelle, *Ammodorcas clarkei*, is found in Ethiopia and Somalia.

The elongated build of the gerenuk is an adaptation to its way of feeding. Like the giraffe, it browses on foliage that grows too high off the ground for most hoofed animals to reach. The giraffe achieves this by being so tall; the gerenuk is a much smaller animal, but is able to stand erect on its long hind legs and stretch its long neck up to browse. It feeds on the leaves and green shoots growing on the upper branches of bushes and small trees, removing them with its mobile lips. Gerenuks are not the only antelopes able to rear up on their hind legs—indeed, all hoofed animals do so from time to time—but they have become highly skilled at both maintaining their balance and walking about while upright. They may eat their way right around a tall shrub without descending to ground level, although they often support themselves by resting their front hooves against the stem.

Since it can get at fresh, juicy leaves that are out of reach of other gazelles, the gerenuk has little use for

ABOVE LEFT **Springbok are always on the lookout for predators. If they sense danger nearby, they race away in the opposite direction, repeatedly breaking their strides with** athletic leaps covering up to 50 ft. in one bound. ABOVE **Stages in springbok courtship: a female nuzzles the face of a male (A), who rubs his facial scent gland against her flank (B).**

tough foliage and coarse, fibrous grasses, leaving these for its less specialized relatives. Many of the plants on which it feeds have succulent leaves that retain water in the hottest, most arid climates, and they satisfy most of its water requirements. The gerenuk rarely needs to drink, and it is able to thrive in semi-desert regions far from water sources. It remains a successful animal in many parts of East Africa, although it continues to decline in the southern part of its range.

Living in the same part of Africa, the dibatag is a gazelle similar in many ways to the gerenuk. Though slightly smaller, it too has a long neck and slender legs. The dibatag follows the same habit of standing erect to

ABOVE A male impala stands guard over his harem of females and young. When alarmed, a herd of impala may leap frantically about in all directions. The dazzling display may serve to confuse predators that are trying to single out an intended victim.

RIGHT In the breeding season impala are aggressively territorial. Here two rival males encounter one another (A); they may retreat without fighting (B), but often they will display head-to-tail (C) before becoming locked in combat (D).

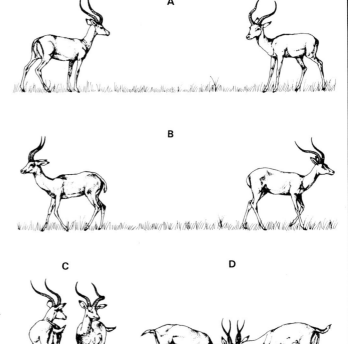

browse, but it has not perfected the technique to quite the same extent as the gerenuk. Its coat is grayish with reddish-brown overtones and a white underside, and it has a long black-tipped tail that is held upright when the animal is alarmed.

The acrobatic springbok

One of the most striking features of the gazelles is their habit of leaping in the air when they are alarmed or excited. All gazelles do this to some extent, but the habit is most highly developed in the springbok. When a herd of springbok are surprised by a predator, they will take flight like other antelopes. But instead of running away in a straightforward manner, they interrupt the sprint with acrobatic leaps, springing 10 ft. or more in the air with their backs hunched up, before landing and continuing their headlong

Gazelles range over Africa, the Middle East, India, central Asia and parts of China.

progress. This extraordinary behavior is called "stotting" or "pronking." Their excited state is emphasized by a line of white hairs that stands up along the hind part of the spine to form a stiff crest; normally this is concealed by a fold of skin. The gazelles' general coloring is bright reddish-fawn with a broad black band on each flank. The underside, rump flash and head are white, and a dark band runs from each eye across the muzzle.

Springbok are gregarious animals, and only adult males show any tendency to live in solitude. When they migrate in search of new pastures—for example when drought has destroyed the food supply—they gather in large herds that, at one time, numbered more than a million individuals. Reports from the late 19th and early 20th century tell of migrating armies of springbok that stopped for nothing, trampling the vegetation in their path.

Such spectacles are now a thing of the past, for springbok numbers have greatly declined in the 20th century. Nevertheless the species is still quite common in some areas of southern Africa, and although its mass migrations are not as impressive as they once were, they are still a reminder of the vast wealth of animal life that existed in the continent before the arrival of the settlers from Europe.

The graceful impala

The impala is a slender, graceful animal with a straight back and long, slim legs. Males may grow to about 3 ft. in height at the shoulders, and the lyre-shaped, ringed horns (present in the male only) may be 20 in. long. The impala's back is a glossy

IMPALA CLASSIFICATION

The impala, *Aepyceros melampus*, is a slender antelope about the size of a fallow deer, found in East, southeastern and southwestern Africa. Though it is gazelle-like in general build, its classification is uncertain. Formerly grouped with the gazelles (the Antilopini), it has been linked with the reedbucks and waterbuck tribe (the Reduncini), and many now consider it to be closely related to the gnus and hartebeests tribe (the Alcelaphini).

reddish brown or golden-tan color, contrasting with lighter flanks and white undersides. It has conspicuous black stripes on its tail and haunches around its white rump flash and on its hind legs.

Impala are agile animals, able to leap to heights of more than 10 ft. and to cover over 30 ft. in one bound. They live on the acacia savannahs or sparsely wooded areas of eastern and southeastern Africa, and parts of southwestern Africa. They are active day and night and, although they have a taste for the foliage of acacia and other shrubs, they will eat whatever vegetation is abundant. Large amounts of grass may be eaten in the rainy season, while their diet consists mainly of leaves in the dry season.

Youth clubs

The sex ratio among impala is weighted in favor of females—only one male is born for every two females. The latter form herds of up to 100 adults and young, accompanied by a small number of mature males. Young and bachelor males live in separate, looser groups of up to 60 individuals. These are known as "youth clubs" and are usually led by the older animals. In the breeding season successful males establish territories, allowing only females and very young males to enter. The females wander from one territory to another, without a permanent link to any one area or any individual male.

Gestation lasts six to seven months, and the newborn young are kept hidden during the first days of life. Mothers and calves form small "nursery herds" which last until the young animals are able to move in a coordinated fashion and can stand on their own; they then rejoin the main herd. In East Africa, young males leave the main herds to join the "youth clubs" when they are six months old; in southern Africa, this happens at one-and-a-half years old.

Easy meat

The impala's main enemies are leopards, cheetahs and Cape hunting dogs. The newborn young are particularly vulnerable, and about half the young born each year fall victim to predators during the first few weeks of life. The predators eat their fill with the sudden glut of young impala and lose interest in hunting. By the time they concentrate on hunting again, the remaining young impala are old enough to avoid capture by running away.